EXPLORING ANIMAL SOCIAL NETWORKS

EXPLORING ANIMAL SOCIAL NETWORKS

DARREN P. CROFT, RICHARD JAMES,
AND JENS KRAUSE

PRINCETON UNIVERSITY PRESS · PRINCETON AND OXFORD

Copyright © 2008 by Princeton University Press
Published by Princeton University Press, 41 William Street,
Princeton, New Jersey 08540
In the United Kingdom: Princeton University Press, 6 Oxford Street, Woodstock,
Oxfordshire OX20 1TW

ISBN: 978-0-691-12751-4
ISBN (pbk.): 978-0-691-12752-1

Library of Congress Control Number: 2008920558

British Library Cataloging-in-Publication Data is available

This book has been composed in Times Roman
Printed on acid-free paper ∞
press.princeton.edu

Printed in the United States of America

10 9 8 7 6 5 4 3 2 1

Contents

Preface

The plan for this book started with a visit of Princeton University Press editor Robert Kirk to Leeds University in the spring of 2005, during which he asked whether we had an idea for a new book. Inspired by a trip to Kenya and its wealth of animal life, we were at first thinking of writing a broad text on how to analyze animal group structure in the form of an edited volume. However, after some thought we came to the conclusion that a more focused book specifically on social network analysis might be more useful (and also avoid overlap with other texts).

In the context of conference talks and departmental seminars, we have frequently been asked by colleagues and students over the last couple of years where to find information on how to use the network approach. However, no text on this topic exists that is written specifically for behavioral biologists. Furthermore, we noticed that many biologists adopt a quasi-networks approach, albeit without statistical analysis. This encouraged and finally convinced us to attempt to write a book on this topic.

One of the problems of writing such a book is that—unlike some areas of biological statistics—this one is far from being a mature subject. Many different views on how to analyze networks abound in the literature, and large numbers of articles on methodological issues are regularly published. The fact that this field of research is developing so rapidly is the reason that we included the word "exploring" in the title of our book. The aim of this book is to introduce novices to the study of social networks and to invite them to join us in their exploration. For the more-advanced reader we hope to provide stimulation partly by drawing attention to some of the more tricky issues involved in social network theory. We readily admit that many problems in this field remain unresolved and that we cannot provide final answers. Nevertheless this should not discourage us from carrying out the many interesting things that already can be done using social network theory and start addressing some of the tougher challenges in this field.

A large part of the planning for this book and some of the writing was done during a retreat at the Ptarmigan Lodge in the Cairngorm National Park in Scotland—a location we can strongly recommend to fellow authors in search of a peaceful place away from the distractions of everyday university life.

Thanks are due to Herbert Krause for his drawings that illustrate this book. We are also greatly indebted to Graeme Ruxton, who read and commented on most of the book and made many useful suggestions, and to David Mawdsley, who worked with us on the network communities. We would also like to thank Jerome Buhl, Seth Bullock, Iain Couzin, Safi Darden, Robin Dunbar, Julia Fischer, Dan Franks, Kevin Laland, David Lusseau, Lesley Morrell, Jason Noble, Andy Sih, Fritz Trillmich, Ashley Ward, Hal Whitehead, and Jochen Wolf for stimulating discussions on various aspects of network theory and application.

Bryan Shorrocks and Tim Clutton-Brock kindly gave us permission to use the giraffe and the red deer data, respectively.

Financial support came from the EPSRC, NERC, and The Leverhulme Trust.

EXPLORING ANIMAL SOCIAL NETWORKS

1

Introduction to Social Networks

Understanding the link between individual behavior and population-level phenomena is a long-standing challenge in ecology and evolutionary biology (Lima and Zollner 1996; Sutherland 1996). Behavior is expressed as a response to intrinsic and extrinsic factors, including an individual's physical and social environment, the latter made up of nonrandom and heterogeneous social interactions (Krause and Ruxton 2002). That is, individuals are part of a network of inter-individual associations that vary in strength, type, and dynamics. The structure of this social network has far-reaching implications for the ecology and evolution of individuals, populations, and species. For example, the social network supports a diverse array of behaviors that will be influenced by its structure, including: finding and choosing a sexual partner, developing and maintaining cooperative relationships, and engaging in foraging and anti-predator behavior. Such behavior is manifested at the population level in the form of, for example, habitat use, disease transmission, information flow, and mating systems, and forms the basis for evolutionary processes including adapting to changing environments, sexual selection, and speciation. Improving our ability to scale up from the individual to the population by establishing why certain patterns of association develop and how inter-individual association patterns affect population-level structure will revolutionize our understanding of the function, evolution, and implications of social organization.

Across the animal kingdom there is immense diversity in social behavior. Social interactions differ in their type (they might be cooperative, antagonistic, or sexual, for example) as well as their frequency and duration; social bonds may last for years or just minutes or seconds. Which type of interaction occurs and with what frequency and duration will depend on factors such as dominance, body size, sex, age, and parasite load of the participating individuals. This raises the question of how we deal with multiple interactions and complex interaction patterns that can arise even if the number of participants is relatively small. Interestingly, sociologists started addressing this question more than sixty years ago when looking at human interaction patterns, and this literature in combination with recent advances in areas such as statistical physics has provided us with a powerful set of tools for the analysis of animal social networks. These tools make it possible to calculate quantitative metrics describing social structure across different scales of organization, from

the individual to the population. The aim of this book is to explore some of
the techniques of network analysis that might be applied to a study of animal
social structure.

1.1 WHAT IS A NETWORK?

The essential elements of a network are "nodes" and "edges." In a graphical representation of a network, each node is represented by a symbol, and
every interaction (of whatever sort) between two nodes is represented by a
line (edge) drawn between them. In the context of a social network, each node
would normally represent an individual animal (though see later in this chapter for some alternate approaches) and each edge would represent some measured social interaction or association. For example, figure 1.1 represents the
social network for a population of bottlenose dolphins, *Tursiops truncatus*, in
New Zealand (Lusseau 2003). In figure 1.1 each filled circle (node) represents
an individual dolphin and the connections (edges) between them indicate a
certain frequency of social contact over a six-year period. This is the type of
network we wish to explore in this book. As we will see as our exploration
continues, much of the quantitative analysis of animal social networks is performed not on a graphical representation of interactions but on a matrix of
values that conveys the same information. Both the graph and the matrix are
representations of the same network.

It should not be a surprise to learn that there are many systems, in many
walks of life, that can be thought of as a collection of pair-wise connections
between objects. Some types of network are very familiar. Probably all of us
regularly tap into a telephone network on which we may simply and quickly
be connected to pretty much anywhere in the world without giving it much
thought. Other technological systems such as electrical power grids (Xu et al.
2004), transport systems (Sen et al. 2003), and the World Wide Web (Tadic
2001) are all quite naturally considered as a network.

Many people have discovered that network theory may provide novel insight into the local and global properties of a system of interconnected objects
that is not possible from considering either the interactions between pairs of
agents in isolation or from a study of the average properties of the system
as a whole. This has lead to researchers studying networks across a range of
systems to gain understanding both of their structure and of some of the consequences of this structure. For example, applications of network theory to
technological systems include optimizing the efficiency of telephone communication systems and analyzing the vulnerability of power grids to the loss of
a power station.

Mathematicians and statistical physicists have made important contributions to the network literature, providing concrete results on the properties of

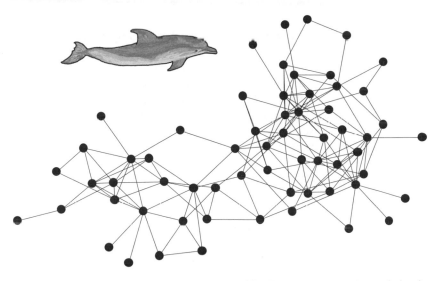

Figure 1.1. Social network of a bottlenose dolphin (*Tursiops truncatus*) population in New Zealand (Lusseau 2003). The network consists of *nodes* (drawn as filled circles here) that denote individuals, and *edges* (straight lines) that symbolize some form of relation. In this network, an edge is drawn between two animals if they were seen together more often than expected by chance. A total of 64 adult dolphins are interconnected by 159 edges.

certain large networks with random assignments of edges to nodes (Erdös and Rényi 1959; Bollobás 1985) and unearthing new paradigms for the characterization of the structure of complex networks and some of the processes that might occur on them (for excellent reviews of the world of networks from a physics perspective, see Albert and Barabási 2002; Newman 2003a; Boccaletti et al. 2006).

The networks approach has also been embraced by biologists interested in unraveling the interplay between cell function and the intricate web of interactions between genes, proteins, and other molecules involved in the regulation of cell activity. They are developing a general framework in which the biological functions of a cell can be understood by examining the structure of its interacting components (Kollmann et al. 2005), enabling them to move beyond "parts lists" of a system and to understand how its components interact to produce complex patterns and behaviors (Jasny and Ray 2003). For example, networks have been used to understand how selective forces have acted on the function of metabolic pathways (Rausher, Miller, and Tiffin 1999) and how gene regulatory networks shape patterns of development (von Dassow et al. 2000; MacCarthy, Seymour, and Pomiankowski 2003). A similar approach has been applied at other levels of organization (Proulx Promislow, and Phil-lips 2005; May 2006). For example, biologists have investigated how cells and organs interact by studying neuronal networks (e.g., Laughlin and Sejnowski 2003), and have considered the structure and stability of ecological systems by plotting trophic interactions between species in the form of a food web (Sole and Montoya 2001; Dunne, Williams, and Martinez 2002). By comparison, relatively few biologists have built and analyzed animal social networks. We will of course be discussing their work throughout the book.

All this wealth of interest from various parties is both a good thing and a potentially bad thing for the budding network analyst who wishes to construct and analyze the structure of an animal social network. On the plus side, there are now many methods and measures that might be brought to bear, and many sources of novel methodology. On the minus side, it must be realized that some of the methods and results derived by one community of researchers do not necessarily translate directly to the analysis of all networks. For example, many of the results obtained by the mathematics and physics community are only applicable for networks with a very large number of nodes. A network such as the internet has so many nodes that its statistical properties may very accurately be approximated by some of the statistical models developed in the physical sciences. Looking for the same properties in a network with a few tens of nodes may, on the other hand, not be so well advised. The type of data that is collected can also have a large effect on how one should go about analyzing a network. Sen et al. (2003) studied the network structure of the Indian railway system. Here the edges represent physical connections (train tracks) between stations, so we can be reasonably confident that the network is an accurate representation of the real system. In contrast, social animals often live in fission–fusion societies (Krause and Ruxton 2002) and their social net-works have to be inferred from observations of interactions been individuals or within groups of individuals. This creates a number of methodological issues associated with the sampling effort required to get a representative picture of the "real" network structure. It is essential that we take such factors into con-sideration when we are exploring and analyzing animal social networks.

1.2 SOCIAL NETWORKS AND RELATED METHODS

Social network theory has its origins in a number of different fields of research on humans. It goes back to the work of psychologists and sociologists in the 1930s who applied elements of mathematical graph theory to human relation-ships (e.g., Moreno 1934; Lewin 1951), and has mostly been concerned with the scenario in which each node represents a single person and each edge some interaction or relationship between two people. A great deal of progress has been made in the analysis and modeling of human social networks in the past twenty or thirty years, made possible by the advent of readily available and cheap computing power, which enables randomization tests and other simulation techniques to calculate more sophisticated measures of social structure and to bring some much needed statistical rigor to the field. The books by Wasserman and Faust (1994), Scott (2000), and Carrington, Scott, and Wasserman (2005) provide an excellent account, from various angles, of many of the methods that have arisen in the social sciences, and we will refer to these sources frequently throughout the book. In recent times social network theory has also received important impulses from the physics community, which have contributed a number of important theoretical advances such as the small-worlds concept (Watts and Strogatz 1998), algorithms for community detection in networks, and qualitatively new insights into the spread of information through populations (Boccaletti et al. 2006).

Network theory provides a formal framework for the study of complex social relationships. Human social networks have been used to investigate a range of topics. These include the spread of HIV (Potterat et al. 2002), the interconnectedness of company boards of directors (Battiston, Weisbuch, and Bonabeau 2003; Battiston and Catanzaro 2004), and the spread of rumors (Moreno, Nekovee, and Pacheco 2004). In contrast to studies on human social networks, the use of network theory to study the social organization of animal groups or populations is still relatively uncommon (Sade et al. 1988; Connor, Heithaus, and Barre 1999; Fewell 2003; Lusseau 2003; Croft, Krause, and James 2004a; Cross et al. 2004; Flack et al. 2006). Perhaps not surprisingly, some of the earliest applications of ideas developed to study human social networks to other animals came in primatology, though such studies generally did not involve statistical validation of the observed patterns (Sade and Dow 1994). More recent studies (Lusseau 2003; Croft, Krause, and James 2004a) have compared quantitative network measures against null models, or used methods inspired by developments in network theory from the mathematics and physics literature (Lusseau and Newman 2004; Wolf et al. 2007) to relate heterogeneities in animal network structure to the biology of their system. However, despite the vast number of studies in the animal behavior literature that have collected information on interactions or associations between pairs of animals, very few investigations have used a network approach to analyze them. We believe that network theory may offer an exciting method to analyze both new and old data sets, which could provide insights into the structure of animal societies not possible with traditional methods.

At this point it seems appropriate to deal with any nagging doubts in the minds of some readers that this is all something you have seen before, just dressed up in new terminology. Surely, you might be thinking, the matrix of pair-wise interactions is nothing more than an association matrix (Whitehead 1997), and its visualization a sociogram (e.g., Zimen 1982; Sade 1989). Don't we already look for collections of closely associated animals in an association matrix using cluster analysis such as Ward's or unweighted pair group method with arithmetic mean (UPGMA) methods (Whitehead 1999)? Well, the answer is a simple yes. An association matrix and a sociogram are indeed different names for a social network. So what is new, then?

The principal advantage, as we see it, of using the network approach to probe the structure of animal societies is that it allows us to tap into a very wide range of measures and approaches that are, as we have hinted, still being developed in parallel in several disciplines, and to apply these all under the umbrella of a single description of the data and the associated analytic tools. Thus we might learn tricks and methods from all manner of sources that might help us unravel what the important structural elements are in our animal social system, and what biological or other factors might be driving that structure. Of course, the real appeal of any approach that amalgamates many interactions is that in principle we can probe structure on all scales from the individual to the population. We then need robust measures that describe the properties of individuals, communities, and populations; our belief is that there are many methods for achieving this that come under the networks umbrella, that just happen to have been developed in the social or physical sciences. In addition, the analysis and visualization of social connections can often be rolled into a single computer program. Network theory therefore offers an "all in one" package that allows us to move between different levels of social complexity, and to tap into new analytic tools.

As we have already mentioned, the vast majority of social networks employ one node to represent a single individual, and each edge represents some form of interaction or association between two individuals (see figure 1.1). Furthermore, each edge in a social network represents the same type of interaction or association. The animal social networks we will analyse for most of this book fall into this category. Many of the systems we will study in this book are "fission–fusion" societies, in which animals frequently leave or join groups (Krause and Ruxton 2002); examples include species of ungulates, primates, cetaceans, fish, and insects. To investigate the fine-scale structure of social networks in such systems, we need to be able to identify individual animals. However, for some species this is not possible (or too time consuming) due to problems associated with identifying, capturing, or recapturing individuals. In such instances it can be useful to identify categories of individuals and consider interactions between them (rather than the individuals themselves). We will illustrate now how these interactions can be considered as a network.

In a study of a captive wolf pack (*Canis lupus*), Zimen (1982) made observations on social interactions in a 6-hectare enclosure over a 10-year period (figure 1.2). During this time juveniles matured into adults and the social status of individuals changed; for example the rank of alpha male was occupied by six different animals. The study looked at a number of different behaviors,

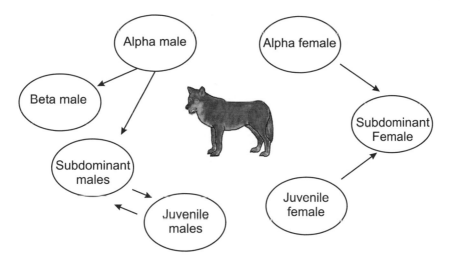

Figure 1.2. Sociogram of "following" events in a captive wolf pack (redrawn from Zimen [1982]). A total of 49 individuals were monitored over a 10-year period during which the social status of individuals changed. The sociogram summarizes the aggressive "following" interactions between individuals belonging to different social categories; arrows point toward the class of animal forced to keep its distance.

one of which was termed "following," which involves a dominant individual forcing a subordinate to keep a certain distance.

Figure 1.2 depicts the "following" behavior in terms of a network where nodes represent classes of animals, not individuals. It clearly shows that following occurs within, but not between, the sexes. Note also that the wolf network in figure 1.2 has directed interactions (i.e., the edges connecting the nodes in the network have arrows that illustrate "who followed whom"). For example juvenile females followed subordinate females but subordinate females did not follow juvenile females, so the interaction is represented with a directed edge going from juvenile females to subordinate females. Based on the information in the network, we could now formulate hypotheses to explore further the social behavior of wolves. For instance, we could ask whether this type of sexually segregated behavior is generally the case with aggressive interactions, or is specific to "following" behavior.

Another alternative approach to network construction that avoids the need for individually recognizable animals is to use nodes to represent behaviors in the population and categorize the individuals that are important for regulating the behaviors as the edges. Fewell (2003) adopted this approach to construct a network (figure 1.3) of the control of pollen foraging in a colony of honeybees (*Apis melifera*). Visualizing behaviors as a network makes it possible to

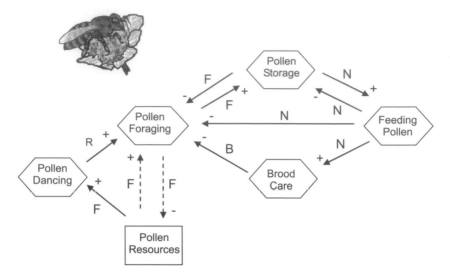

Figure 1.3. A process-oriented network depicting pollen-foraging behavior in the honeybee (redrawn from Fewell [2003]). In this network the nodes represent tasks within the colony and edges between them are the individuals that transmit the information: F, forager; N, nurse; B, brood; R, recruits. Both positive (+) and negative (−) feedback are indicated.

investigate how each caste in the colony contributes to overall system function. Fewell's (2003) network explains the modulation of foraging behavior within the colony. As foragers place pollen in the pollen cells, they receive information on the amount of pollen stored, which feeds back negatively on pollen foraging (i.e., the less pollen stored, the more foraging occurs). Nurse bees remove the pollen from the cells and feed it to the developing brood, and give access to the pollen foragers. Receiving access to pollen from a nurse bee feeds back negatively on foraging behavior (i.e., the more feeding, the less foraging occurs). When hunger levels in the brood are high they produce a hunger pheromone that encourages foraging behavior; brood care reduces hunger levels. Finally pollen dancers provide information on pollen availability and location, and dancing also elicits recruitment to foraging by workers that are not actively engaged in foraging.

1.3 OUR MOTIVATION FOR WRITING THIS BOOK

There are several reasons why we think it is timely to write a book exploring networks from the perspective of researchers working with groups of individually recognizable animals. First and foremost, we believe that a networks approach has great potential as a means to perform a range of quantitative analyses on animal social structure on all levels from individuals to populations. As we will demonstrate in the following chapters, a networks approach allows us to assign quantitative measures of social structure to individuals and populations. These measures open up exciting opportunities for data analysis. For example, they can be analyzed in the context of measured attributes of individuals such as morphology (e.g., body size), behavior (boldness, say), or reproductive success, plus inter-individual measures such as relatedness, all of which will help us shed new light on the mechanisms and functions underpinning, and underpinned by, animal social structure. The fundamental point is that understanding network structure will potentially tell us substantially more about the individual and population than will information on individual attributes alone or interactions between individuals in isolation.

One of the messages that we hope will ring clearly from this book is that animal social network analysis is a work in progress. Much of what we might think of doing is still beyond us, but a second aim of this book is to stimulate interest in developing the remaining tools needed to make the approach an indispensable tool in the behavioral sciences. Before we list some of the potential uses of networks, let us whet the appetite with an illustration of the sort of thing that can already be done. To this end we made up a data set for an imaginary species (*Commenticius perfectus*) that we hope illustrates a number of useful techniques. In the logbook, or even loaded into a database or spreadsheet, a relational data set can look like an impenetrable tangle of numbers, with no discernable pattern to it. One of the first steps (and often a rather major one) in making sense of network data is to plot it. As we will explore more fully in chapter 3, the way we plot it can make a big difference.

Figure 1.4a shows that even for our very simple example, with only 20 nodes (animals) and 35 edges (interactions or associations) a network can look just as featureless as the original data. However, many computer packages exist that can very easily be instructed to lay out the nodes and edges in a way that often looks much more appealing. Figure 1.4b shows the same network, but laid out using a technique called spring embedding (see chapter 3). From this it is immediately clear that all animals are somehow interconnected into a single network, but that not all animals are equal in terms of the number of connections they have, for example, or in terms of whether they occupy central or peripheral positions in the overall network.

In our experience, this first simple visualization of hard-won relational data often has a large impact on the researcher, and almost invariably induces interesting questions and hypotheses about why the network structure is the way it is. We can often help ourselves by adding individual identities and attribute data such as sex or body size (fig. 1.4c). This information helps us to understand how phenotypic attributes influence who is connected to whom in the network and which individuals are central and which peripheral.

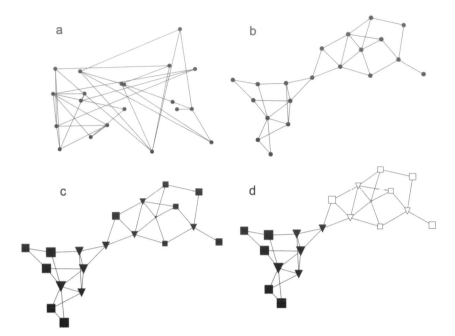

Figure 1.4. Four visualizations of the same animal social network for a fictitious species, *Commenticius perfectus*: (a) random layout, (b) spring embedding, (c) with node shape and size depicting animal attributes: shape indicates sex (square = male) and size indicates body length, (d) with node shading indicating community membership (the other node features [size and shape] are as in c).

Once we start to pose questions of our network, then we need to quantify its structure. We can calculate a whole range of descriptive statistics that measure the local and global properties of the structure of our network, including measures of the overall network size, how interconnected the local network neighborhoods are, how interconnected the network is globally, how central or peripheral different classes of individuals are (e.g., male versus female), and so on. A detailed explanation of some of these measures can be found in chapter 4. We can also look for substructures—so-called communities—in our network and we can identify those individuals that interconnect communities and relate their key position in the network to some of their attributes (fig. 1.4d). Finally, we may want to compare the structure of our *C. perfectus* network with other networks, perhaps for the same population under different environmental conditions or based on different behavioral interactions, or indeed with other populations or species, to gain insight into the generality of the observed patterns or the influence of external factors (such as the environment) on the network of social interactions. An exploration of the options for comparing networks can be found in chapter 7.

So what biological questions might we be able to address with a networks approach? We believe that there are many, but here is a brief discussion of a few of the possible avenues of research.

Social network analysis allows us to focus on individuals and examine the influence of the network on individual behavior. The notion that the social environment has fitness consequences for the expression of individual behavior is central to game theory (Maynard Smith 1982), and theoretical work on behavioral strategies has demonstrated the importance of interaction patterns (Nowak and May 1992; Nowak, Bonhoeffer, and May 1994). The network approach puts individual behavior in the context of the population social structure, thereby helping us to understand the evolution of behavioral strategies. For example, Ohtsuki et al. (2006) showed that the evolution of cooperation is strongly dependent on the fine-structure of social networks. They found that in networks representing perfect social mixing (i.e., each individual is equally likely to interact with each of the others), defectors benefit from exploiting cooperative individuals. However, if individuals do not live in a perfectly mixed world, then selection may favor cooperators when the average number of connections per individual is sufficiently small. In other words, the network structure may determine whether a cooperative strategy is able to persist in a population. The network approach thus puts individual behavior in the context of the population, thereby helping us to understand the evolution of behavioral strategies.

It has been recognized for some time that considering the social environment in which behavior is expressed is important in the context of signaling (McGregor and Dabelsteen 1996). There is a growing literature on "communication networks" that connect an individual signaler to each of its receivers. The approach has provided important insight into the design and function of signals and the role of the social environment on signal evolution. These communication networks are defined for each individual under investigation. There is obviously great potential in coupling a social networks approach with that of communication networks. One area that may be particularly productive is to look at the transmission of information across a social network spanning multiple communication networks (and which therefore contains individuals that are not in direct communication range).

A networks approach also has the potential to illuminate the influence of individual behavior on network structure and function. For example, Flack et al. 2006 studied the construction of social niches in primates. Using "knockout" experiments on a captive group of pigtailed macaques (*Macaca nemestrina*), they demonstrated the importance of a small number of individuals in the population that performed policing behavior (intervention during conflicts). When the policing individuals were absent from the network, social niches destabilized, with group members forming smaller, less-diverse and less-interconnected networks across a range of behaviors including play and sitting in direct contact. There is obviously great appeal in extending this approach across different taxa and behaviors.

Probably the main strength of the network approach is its potential to address population-level or cross-population-level problems by building up complex social structures from individual-level interactions. Thus the network approach bridges the gap between the individual and population, and this is highly relevant to modeling population-level processes. For example, a network approach allows us to identify who is connected to whom in the population, information that helps formulate hypotheses as to who will learn from whom (Latora and Marchiori 2001), or who will infect whom with a disease (Watts and Strogatz 1998; Corner, Pfeiffer, and Morris 2003; Cross et al. 2004). An understanding of the fine-scale interaction patterns between individuals revealed by social networks is a major advance on the assumption of random interactions by traditional epidemiological models of disease transmission or models of social learning. In a similar vein, we now understand that many social systems contain elements of self-organization in that behavioral interactions strongly influence population-level processes that in turn will feed back on the individual (Camazine et al. 2001). It would seem likely that a networks approach can help to make both parts of the process, individual→population and population→ individual, more tractable.

Another motivation for writing this book is that we believe that there are plenty of behavioral scientists who are using a graphical network approach, but not following it up with quantitative analysis, and others who have relational data sets which could fruitfully be analyzed using network methods. We would like to encourage anyone in either camp to have a go. It is remarkable how commonly behavioral biologists use a quasi-network approach to their study system. At a recent conference on animal behavior, we found that 25 percent of the posters showed data in the form of networks. Interconnecting individual animals with lines is a popular approach in studies on birds and mammals (particularly primates; see figure 1.5 for an example). However, often the authors appear not to be fully aware of the power of the network approach and the wealth of information that can be extracted from their data set in this way, provided the data were collected in an appropriate form (see chapter 2). Frequently networks are used purely as a graphics tool but not an analysis tool. This discrepancy between the widespread use of a network representation and the lack of, or superficial use of, statistical analysis is what prompted us to write this book.

We aim in writing this book to offer a synthesis of some of the very large networks literature that is most appropriate for behavioral biologists. Despite the wealth of literature on networks in the field of sociology and the physical and mathematical sciences, we felt that a book devoted to the needs of biologists is merited for a number of reasons. Many of the publications on network theory in the physics and mathematics literature are highly technical

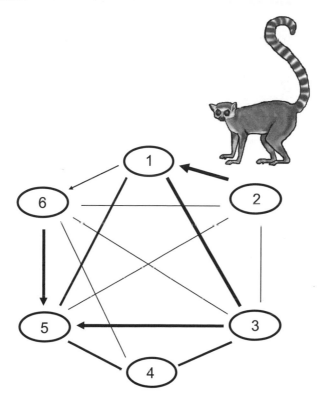

Figure 1.5. Sociogram of infant licking among mothers in ring-tailed lemurs (*Lemur catta*) (redrawn from Nakamichi and Koyama [2000]). Line thickness indicates frequency of interaction and arrows indicate nonsymmetric relationships (when one mother licking another mother's infant occurs significantly more often than expected by chance).

and difficult to grasp for the uninitiated reader. We have tried to provide an introduction to the concepts and tools of network theory suitable for novices of the subject. Though some of the literature in the social sciences is more digestible, there are differences in approach that must be addressed. Biologists in the behavioral sciences are usually interested not only in the mechanism but also the function and evolutionary origin of behavior. For instance, a biologist might want to understand what selection pressures have led to the evolution of particular network structures (i.e., interaction patterns between individuals). A comparative study of networks of different populations of the same (or closely related) species might be one way of addressing such a question. Such an approach can lead to the development of questions fundamentally different from those of interest to sociologists and psychologists. Furthermore, the issues involved in experimental design and data collection—a topic to which chapter 2 of this book is dedicated—differ considerably between sociology and behavioral biology. It is crucial for behavioral biologists to replicate their experimental trials to test for the generality of the observed patterns and ideally to manipulate their system experimentally to investigate the underlying mechanisms and functions. We must also be aware that our sampling regime might influence network structure—a problem that we will return to at various points in later chapters. It is our hope that this book will go some way toward helping colleagues address these problems in future studies on animal social networks.

Our final aim is to apply the occasional "over-enthusiasm filter." Some of the physics literature on networks is concerned with data sets much larger than anything a behavioral biologist is likely to encounter and thus subject to different statistical approximations. We have tried to provide the reader with network theory that is "ready to use" for sample sizes that biologists are likely to work with. To illustrate this point, many if not most biologists collecting data on social networks are unlikely to obtain data sets large enough to facilitate a realistic test of scale-free properties of their system (see chapter 4 for details). The latter is a topic of intense interest among physicists and mathematicians (see Barabasi and Bonabeau 2003) among others, but unlikely to become a major issue for most behavioral biologists who observe and collect information on a few tens or hundreds of primates, cetaceans, or ungulates.

1.4 AN OUTLINE OF THIS BOOK

This book is aimed primarily at behavioral biologists who wish to collect data on the social organization of animal groups or populations in captivity or in the field. It is meant to give practical advice on how to design a study and how to collect, present, analyze, and interpret data using a social network approach. Although each section and each chapter should allow the reader to complete a certain analytical process that is useful in its own right, we have designed the book so that topics of increasing complexity are covered. Some readers may feel that all they need is some information on descriptive statistics. Others may be keen to run simulations and use or further develop algorithms that detect community structures. Therefore we hope that this book has something to offer to beginners as well as to established scientists in this field of research who share our interest in and fascination with social networks.

Wherever possible we will point the reader to one of a number of software packages that facilitate the calculation of descriptive network statistics and the running of tests and simulations (see box 1.1). These packages make the network approach more accessible to a wide audience. Though the book is not a guide to using a particular analysis or graphics package, we have made considerable use of UCINET and NETDRAW to illustrate the text. From our experience both programs are easy to use and suitable for exploring networks in animals. Free trial versions can be obtained for both programs via the Internet. There are many other software packages out there that will do the same job (see Huisman and van Duijn [2005] for a review).

Box 1.1

An Overview of the Programs Referred to in the Book

In a recent book, Huisman and van Duijn (2005) provided a comprehensive review of both commercial and free packages for analyzing social networks. The appendix in Scott (2000) also contains a summary of many packages. To illustrate the potential of social network analysis, we have used UCINET (Borgatti, Everett, and Freeman 2002), which is probably one of the more frequently used software package for the analysis of human social network data (Huisman and van Duijn 2005). Other programs and packages have also proved very useful. Here we present an overview of the main programs featured in this book.

UCINET is a comprehensive package for the analysis of social networks (Borgatti, Everett, and Freeman 2002). It can read and write to a range of different format text files in addition to Excel files. It is able to deal with very large data sets—networks that contain up to 32,767 nodes (individuals), far more than most of us are ever likely to be able to collect data on. UCINET offers a range of network analysis methods and procedures, including many of the techniques described in the following chapters in this book. Overall UCINET is a very useful and user-friendly package. Integrated with UCINET is the NETDRAW program (see below) for drawing social networks.

NETDRAW is a free program written by Steve Borgatti for visualizing social network data in 2D space. The visualizations are very flexible, and use different algorithms to display the network in a range of formats. It can handle multiple relations at the same time, and node attributes can be used to set colors, shapes, and sizes of nodes. NETDRAW also has some analysis procedures, the most useful of which is perhaps the possibility to visualize communities (see chapter 6) in the network. The network images can be saved in a range of formats including metafile, jpg, gif, and bitmap. There are a number of options for importing files into NETDRAW.

SOCPROG is a series of MATLAB programs written by Hal Whitehead for analyzing data on the social structure, population structure, and movements of identified individuals. SOCPROG works through a user interface, and most procedures can be executed at the click of a button without knowledge of MATLAB. For those not running MATLAB on their computer, there is also a compiled version that will run without MATLAB.

POPTOOLS is a set of tools for analysis of matrix population models, simulation of stochastic processes, and calculation of Monte Carlo and bootstrap statistics in Excel 97 or above. The interface for the program is self-explanatory, and POPTOOLS also comes with worked demonstrations that are extremely useful. If the matrix exceeds the size limit allowed by Excel (255×255 cells, due to the limit on the number of columns in a worksheet), it is possible to use tab-delimited text files to input the distance matrices (more information can be obtained from the help files in POPTOOLS).

To help you find your way through the book we have outlined a number of steps that might be followed when taking a network approach:

1. Formulate an initial question: All scientific work starts with a question for which we want an answer. Whether your particular question is amenable to the use of a network approach depends on a number of criteria. You must be able to recognize individuals (see point 2 below) and observe interactions between them. In most cases you will have to make repeated observations on the same individuals to build up information on interaction patterns within the group (chapter 2).

2. Determine a method for identifying individuals (chapter 2): The construction of networks is usually dependent on the identification of individuals or at least categories of individuals (e.g., castes in social insects). A number of techniques for marking individuals are discussed in chapter 2.

3. Choose a measure of interaction and a research design (chapter 2): The choice of measure for interactions will dependent on the type of investigation. Social interactions between individuals often involve multiple sensory channels (e.g., visual, acoustic, mechanical stimuli). You also need to decide on the number of individuals that you want to monitor and how often and for how long to observe them.

4. Define each interaction measure (chapter 2): What constitutes an interaction and the precise nature of the interaction requires careful observation and definition to standardize the data set and make it reproducible. In many cases co membership in a group, or some other measure of association, may be a perfectly useable proxy for a pair-wise interaction.

5. Select the appropriate recording methods (chapter 2): Observations can be made in various ways (e.g., continuous observation, point sampling, event sampling), and the choice made can have important implications for the quantity and quality of data collected. Field studies are often constrained by animal movements (movement to and from the study site) and mortality, and need to be carefully designed from the outset to make recording representative social networks possible.

6. Organize the data (chapter 2): Information on social interactions between individuals needs to be organized into a matrix for data analysis.

7. Consider sample size (chapter 3): The amount of data that need to be recorded for a study will obviously depend on the question that we want to answer and on how dynamic the study system is. For example, if interactions between individuals are relatively stable, we may be able to get at the social structure with relatively few observations.

8. Construct and visualize the social network (chapter 3): It is often helpful to run a pilot study to test whether the experimental design and data-recording techniques produce the expected results. For this purpose it is useful to monitor how the data set for the network builds up over time and to run some preliminary analyses. A pilot study will usually show very quickly whether for a given sample size a realistic number of repeated observations on individuals can be made that results in meaningful data before the entire study is completed. Looking at the networks can be very helpful in this context. This is the place to jump in if you already have some data and want to use network theory to analyze them.

9. Perform detailed network analysis (chapters 4–7): A number of quantitative metrics can be calculated that describe social structure across different scales of organization, from the individual to the population. Most of the descriptive statistics (chapter 4) can be computed relatively quickly within computer packages. More advanced techniques require statistical tests, some but not all of which are available through computer packages (chapters 5–7).

10. Interpret network measures (chapter 5): Often it is very helpful to compare the observed network to a randomized network that provides a null hypothesis. There are a number of different randomization techniques that need to be carefully distinguished because the choice has an important influence on the results and interpretation of our observed data.

11. Search for sub-structures (chapter 6): A closer look at the fine-structure of networks can help identify subunits (so-called communities) and individuals interconnecting sub-units. Both types of information can be very useful in for-mulating testable hypotheses.

12. Compare networks (chapter 7): Comparing the same set of individu-als and the interaction patterns that interconnect them under different ecologi-cal conditions can provide important information on the social organization of a population. Likewise we can compare the interaction patterns of closely related species or different populations of the same species that are exposed to different ecological conditions. This type of analysis may provide us with insights into the evolution of social organization.

2

Data Collection

To get us started in our exploration of animal social networks, we will need some data. In this chapter we will outline some of the different types of data from which such networks can be constructed, and how these data might be collected. A network is a relational data set in that it represents how each animal may be related to each of the others through the type of interaction or association being observed. The key element in a network data set is therefore the set of pair-wise relations between animals, each of which is represented by a network edge. These relational data are not the whole story, though. Throughout the book, a common theme will be that we will try to interpret the structure of social relations, as revealed in the social network, using properties of each individual as possible explanatory variables. These properties (so-called attribute data) must also be collected and analyzed.

Having outlined what data is appropriate for social network analysis, we will introduce you to the different methods for collecting such data. One of the prerequisites for constructing social networks of the type we are interested in is that individuals can be identified and interactions between them observed. We will provide an outline of some of the techniques available for this and discuss their advantages and limitations. The sampling protocol we adopt will depend on the study system and the definition of a social relation. For example, in the laboratory it may be possible to follow and record every pair-wise interaction, and thereby achieve continuous sampling. In wild populations this will rarely be possible, and we are more likely to be forced to adopt a strategy of point sampling (where we sample the population at discrete time intervals; see Martin and Bateson [2007] for details). If the population we plan to study is very small, it may be possible to mark and monitor all individuals, giving us potentially complete information on the social structure. Of course, most natural populations are too large to allow this, and we will need to decide how many individuals to include and over what time period to sample.

Finally, we will need to present our data in a form that is suitable for subsequent network analysis. The convention is to arrange the relational data into a matrix, and we will discuss the different types of matrices and their use in social network analysis, and give some pointers on how you might arrange your data so as to ease the pain of loading it into appropriate analysis and visualization software.

2.1 RELATIONAL DATA

Social structure can be considered to be the result of behavioral interactions, such as grooming, mating, and fighting, between individuals (Hinde 1976). The essential element of an animal social network is that is represents a relational data set, which encapsulates some form of interaction between individuals. The animals involved in an interaction are always considered "two at a time." That is, the building block of the network, an edge, always connects just two individuals. The interest arises once we have accumulated all of our "pair-wise" interactions to see whether, in fact, our interaction somehow links together most or all of the members of our study system. The overall network of relations, such as those in figure 1.1, describe "who is connected to whom" and how closely. Social network analysis is predominately concerned with the statistical measures of network structure of pair-wise relations (see chapters 4–7).

Generally speaking, there are two classes of relational data that we might consider as the basis for an animal social network (Whitehead and Dufault, 1999). First, we can define pair-wise relations based on *associations* between individuals. For example, individuals that are in the same social group, roost, or nest may be considered to be associating. This is a very common method of compiling social network data, as we shall see. Second, and in some ways preferably, we can draw edges based on an observed behavioral *interaction* between two individuals. There is a wide range of pair-wise interactions that might form the basis of a relational data set; they could be competitive or cooperative interactions, grooming or mating, aggressive or submissive, to name but a few examples.

Happily, both association-based and interaction-based data can be analyzed as networks, though there are some methodological issues that arise in association-based networks that needn't worry researchers collecting genuine pair-wise interaction data. The details of all this can wait until later in the book. However, we have tried to distinguish between the two when it is important. When the ideas we are representing are appropriate to both types of data, we will either refer to "interactions or associations," or use the term "relations" to represent both types of relational data.

Though we will frequently try to explain our network in terms of attribute data, this does not preclude our analyzing the relational data themselves. It is possible (and sometimes useful) to make comparisons within a relational data set using standard statistical techniques. For example, one could compare the number of grooming events that were reciprocated in comparison to those that were not reciprocated (e.g., Hart and Hart 1992; Manson et al. 2004). Of course, we may also want to collect data on more than one type of interaction within the population, to allow us to compare relationships between different interactions. For instance we may be interested to test the hypothesis that animals that form reciprocal cooperative interactions will also frequently socialize together (see chapter 7). The norm is for a given social network to contain edges all representing the same type of interaction. Then to compare interactions we will need to compare separate networks; we will explore how we can do this statistically in chapter 7.

Defining Social Associations Based on Spatial Proximity

While social networks constructed from behavioral interactions will provide detailed information on the nature of social relationships between individuals, such interactions are often difficult to observe. To sidestep this problem many authors have quantified patterns of associations between individuals based on some measure of spatial proximity (e.g., individuals that were observed within the same group) and used these measures of association to describe social structure. (Lest we have given the impression that association-based relational data are inferior, it should be pointed out that in many cases they may be able to reveal social structure that would be very hard to find in interaction data.)

The key question is how do we decide if two animals are associated? A common method is to base the decision on spatial proximity. Then there are essentially two approaches. First, we can use group membership; all individuals in a group are said to be associated. Second, we can base our definition on space use, so that all individuals within the same patch of habitat are said to be associated. Defining the spatial scale over which interactions occur is an important decision in collecting association data. If the spatial scale is too great, sampling will combine individuals into groups where no biologically meaningful association occurs. At the other extreme, if the scale is too small it may exclude important associations. The definition of the spatial scale for associations will depend on the question of interest. For example, if we are interested in the social transmission of information through a network, then two individuals may be defined as associated if they have the potential to exchange information (Bradbury and Vehrencamp 1998). Thus, understanding the sensory channels and their constraints for the communication systems of the study species is of great importance. In contrast, if we are interested in the transmission of a disease that is only transmissible by physical contact (many skin infections for instance), then we may only be interested in animals that have direct physical contact.

Defining Associations Based on Group Membership

Perhaps the simplest way to define social association patterns between individuals is via group membership (recording individuals that occur in the same social group as having a social tie). Such a definition is also of general interest because group living forms the basis for much other behavior, including cooperation, social learning, and reproduction (Wilson 1975; Krause and

Ruxton 2002). Defining associations this way has been dubbed "the gambit of the group" (Whitehead and Dufault 1999), whether the grouping is defined spatially (as is more usual) or temporally. In network terms, invoking the gambit of the group means placing an edge between every pair of animals found in the same group. The gambit has been used to define associations across a range of taxa including: fish (Helfman 1984; Ward et al. 2002; Croft, Krause, and James 2004a), cetaceans (Slooten, Dawson, and Whitehead 1993; Chilvers and Corkeron 2002; Ottensmeyer and Whitehead 2003), and ungulates (Clutton-Brock, Guiness, and Albon 1982; Cross et al. 2004; Cross, Lloyd-Smith, and Getz 2005). We will give quite a lot of attention to analyzing networks constructed this way in the remainder of the book.

When defining associations based on group membership, a number of important issues have to be taken into consideration. First, we need to think about why animals are associating; does the association represent a (potential) social relationship or are the individuals grouping for some other reason? For example, individuals may be associating because of a concentrated resource, such as a food patch, and not for social reasons. Such groups are generally referred to as aggregations (Krause and Ruxton 2002). If we are interested in "who associates with whom" in a social context, then basing our relational data on foraging aggregations may be misleading, as it may not reflect the true social structure of the population. However, if our focus is on the transmission of a disease that requires close proximity of infected and susceptible animals, then information on "who forages with whom" or "who nests with whom" may be appropriate (see the section on defining associations based on space use below). Thus it is important to distinguish between aggregations and social groups (see Krause and Ruxton 2002), and to decide which definition is most appropriate for the study system and the question at hand.

If we are to use "the gambit of the group" to define relations, then we must also decide under what criteria two individuals are considered to be in the same group. A standard method is to base this decision on inter-individual distances. A useful way of estimating distances is in terms of body lengths. If the approximate size of the animal is known, then the number of body lengths between individuals can be converted to standard units of measurement. A sensible method is to employ a "chain rule," as illustrated in figure 2.1. Some threshold distance d is chosen below which an animal and its nearest neighbor are taken to be in the same group; if the nearest neighbor is farther apart than d, the two animals are in different groups.

Of course, the value of d has to be biologically meaningful, and should be informed through quantitative measurements. For example, in an investigation of the red deer (*Cervus elaphus*) on the island of Rum off the west coast of Scotland, Clutton-Brock, Guiness, and Albon (1982) estimated the value of d by sweeping through a group of animals in one direction and estimating by eye the distance from each individual to its nearest neighbor (defined as

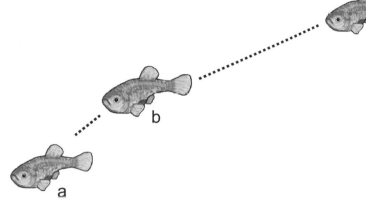

Figure 2.1. An example of group definition based on the "chain rule." Provided the distances from *a* to *b* and from *b* to *c* are less than the threshold distance (*d*), all three individuals are taken to be in the same group, even if the distance from *a* to *c* is greater than *d*.

the individual with its head closest to the focal individual). Clutton-Brock and his colleagues observed a bimodal distribution of nearest-neighbor distances, with most either within 40 m or more than 60 m away. Thus a nearest-neighbor distance of 50 m was used to define individuals within the same group. Further observations confirmed the validity of this definition, with behavioral synchrony occurring between 90 percent of deer in the same group, whereas only 56 percent of deer in different groups had their behavior synchronized.

In investigations on shoaling fish, a criterion of 4 body lengths is often applied (Pitcher 1983; Krause and Ruxton 2002), which is well within the average distance observed between fish while shoaling. In figure 2.1, fish *a* is within 4 body lengths of fish *b*, and *b* within 4 body lengths of fish *c*. Under the chain rule, all three fish are in the same group, despite fish *a* and *c* being more than 4 body lengths apart. Great care must be taken when applying such a chain rule, particularly if "association" is being used as a proxy for "interaction" or, as is more common, "potential interaction" among group members. We must therefore ask ourselves whether all individuals in the group truly could be interacting or associating. In an investigation on guppies (*Poecilia reticulata*), a small freshwater fish, we have used the chain rule to define social associations where all fish observed in the same shoal were deemed to have a direct network connection. This assumption is based on the fact that guppy shoals are sufficiently small (Croft et al. 2003) to allow all individuals in a shoal to directly interact, and the evidence of Griffiths and Magurran (1998), who showed that wild female guppies were "familiar" with randomly selected shoal companions. However, if the investigation were focused on pelagic marine fish shoals or migrating herds of ungulates on the plains of Africa, then such a definition might be inappropriate, as the probability of any two individuals in a group of thousands or even hundreds of thousands interacting is minimal. In such circumstances progress could be made by investigating sub-structuring within large groups. For example, by defining social relations based on nearest-neighbor distances (e.g., Sibbald et al. 2005), it may be possible to investigate the fine-scale social structure within large groups of animals (see section on directed network interactions later in this chapter). This is only going to be possible if the relative spatial positions of the members of a group are known; if they can't be observed, there is little that can be done to increase the spatial resolution beyond that of the group. One course of action available to us in this situation, which is explored in chapters 3 and 5, is to restrict our attention to only those associations that occur frequently, in the expectation that this approach will filter out associations due to coincidental spatial proximity rather than genuine (or potential) social interaction.

Defining Associations Based on Space Use

In a number of studies, shared space use has been used as the basis for defining social association. As we have already mentioned, this might be an appropriate approach for investigations of the transmission of innovations or disease. For example, in an investigation of brushtail possums (*Trichosurus vulpecula*), Corner, Pfeiffer, and Morris (2003) constructed an association network based on which "den" individual possums slept in during the day over a number of successive days. In this study Corner and coworkers were interested in the transmission of *Mycrobacterium bovis* (tuberculosis), for which den sharing provides the greatest risk of transmission. Similar definitions of association have been used to investigate the social organization of Spix's disc-winged bats (*Thyroptera tricolor*) that share roosting sites (Vonhof, Whitehead, and Fenton 2004). Similar ideas could be applied to populations that share feeding sites, watering holes, breeding arenas, and so on.

Defining Network Connections Based on Behavioral Interactions

A more direct approach to constructing animal social networks is possible if inter-individual behaviors, such as competitive, cooperative, mutualistic, and sexual interactions, can be observed directly, rather than inferred from group co-membership or shared space use. As an example of what we have in mind, Sade (1972) collected data on the patterns of grooming behavior (a behavior important for reinforcing social bonds) by recording "who groomed whom" in a population of Rhesus monkeys (*Macaca mulatta*).

It is essential in these cases that the definition of a behavioral interaction is informed by the natural behavior of the animals and the question of interest. In their influential book on measuring behavior, Martin and Bateson (2007) outlined four measurements of behavior: latency, frequency, duration, and intensity. Social networks are generally constructed from information on the frequency or intensity of interactions. This can however, be defined in a number of ways, and different criteria may be used to define a behavioral interaction. If we take interactions defined by reciprocal grooming, for instance, we may use a number of criteria: the reciprocal act may have to be conducted within a certain time frame (latency), or be of equal or longer duration and of similar intensity (e.g., the area groomed) to the initial act. Thus when observing behavioral interactions, it may be necessary to collect multiple measures, and we refer the reader to Martin and Bateson (2007) for a detailed discussion of these issues.

Representing Relational Data

The most usual way to represent and manipulate relational data for social network analysis is to construct an "association matrix" (table 2.1). There are a number of conventions in the treatment of matrices, and we will digress briefly to summarize some of them. An $m \times n$ matrix is a table with m rows and n columns. If there are n animals in our network, its association matrix is an $n \times n$ matrix with each row and each column representing a different individual, with the individuals arranged in the same order along the rows and columns. The whole matrix is often written in bold as **X**. Individual entries (or "elements") in the matrix are written as X_{ij}; the first subscript denotes the row, and the second the column representing the individuals of interest. Thus X_{ij} is the value of the interaction or association between the ith individual and jth individual (table 2.1). (Note that we will not differentiate between matrices derived from associations or more direct behavioral interactions; both will give rise to an "association matrix.")

The simplest form of associations or interactions are undirected and binary. They are binary (or "unweighted") in the sense that each entries in the association matrix is either 1 if the two animals concerned interacted with each other, or 0 if they did not. The interaction is undirected if individual A interacting with B automatically means that B is interacting with A (see table 2.1). This will always be the case if the network is constructed via the gambit of the group, for example. The elements in the association matrix running diagonally from the top left to the bottom right of the matrix represent the direct social relation between any one individual and itself. These are generally left blank because individuals will not normally interact with themselves. There may be exceptions to this; for example, if we are interested in the frequency of grooming events, then we might be interested in the frequency (or occurrence) of

TABLE 2.1.
An example of an association matrix for
5 individuals, labeled 1–5 above each column
and next to each row. A "1" in row 3, column 4
indicates a relation between individuals 3 and 4.

		Individual				
		1	2	3	4	5
Individual	1		1	0	1	1
	2	1		0	0	1
	3	0	0		1	0
	4	1	0	1		0
	5	1	1	0	0	

TABLE 2.2.
Levels of measurement in relational data. Adapted
from Scott (2000).

		Directionality	
		Undirected	Directed
Numeration	Binary	1	3
	Weighted	2	4

self-grooming, which could be represented in the diagonal elements. We may want to consider other features of interactions between individuals and define more complex forms of interaction. Thus, we can classify four ways in which data can be presented for social network analysis, and two variables that can be manipulated: (1) data are binary or weighted, and (2) data are directed or undirected (see table 2.2).

Weighted Network Relations

We can increase the amount of information in the network data set by considering weighted (but still undirected) relations between individuals. In such a data set we are interested not only in the occurrence of relations but their frequency

TABLE 2.3.
An example of an undirected
association matrix for three individuals
with weighted relations. If the data
were association-based, this might
represent that individuals 1 and 2 were
seen together 10 times, 2 and 3 five
times and 1 and 3 on one occasion.

		Individual		
		1	2	3
Individual	1		10	1
	2	10		5
	3	1	5	

or strength or sign. Weighting in the network can simply be represented by coding the nonzero elements in an association matrix with a range of values (see table 2.3). For example, Croft, Krause, and James (2004a) used the number of times two fish occurred in the same shoal over a seven-day sampling period as an indication of the strength of the social interaction between the two individuals. Including weighting into relations allows the network to be filtered to represent its core, nonrandom components, which for some data sets may provide a more detailed insight into the true population structure (see chapters 3 and 5 for details on filtering networks).

Directed Network Relations

Network relations (edges) based either on associations or behavioral interactions may be directed. To see what we mean by this, consider a grooming network. If individual A grooms individual B, but B does not groom A, then the interaction is directed (from A to B) and in the grooming network we would have a directed edge connecting A to B but not B to A. Of course, if B does groom A as well, that particular network connection becomes symmetrical (at least in terms of the existence of the edge—if edge weight is also considered, the symmetry will not be perfect if, for example, A grooms B more than vice versa). Directionality in behavioral interactions can also be a result of who initiated an interaction or due to winner–loser effects during aggressive interactions. Association-based data can also lead to directed networks; for example, if associations are recorded on the basis of considering which animal is another's nearest neighbor, this is not necessarily a reciprocated relationship, as can be seen in figure 2.2.

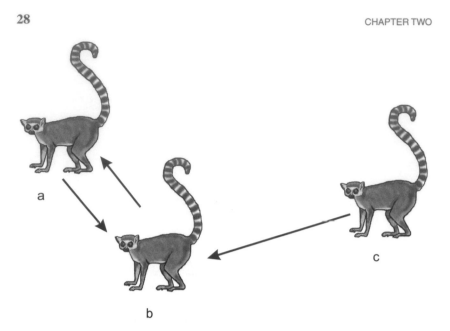

Figure 2.2. An illustration of how a using a nearest-neighbor rule to define social association can lead to directed edges. The nearest neighbor of individual *c* is individual *b*, but this is not reciprocated, as the nearest neighbor of individual *b* is *a*.

TABLE 2.4.
An example of a directed association matrix for three individuals.

		Receiver		
		1	2	3
Actor	1		1	0
	2	0		1
	3	1	0	

An undirected relational data set gives rise to an association matrix (such as table 2.1 or table 2.3) in which the two halves bisected by the "self-interaction" diagonal are mirror images of each other. Any directionality in the pair-wise relation breaks this symmetry. By convention the "actor" in a directed relation is denoted by the row and the "receiver" by the column in an association matrix. For example, in the matrix in table 2.4 actor 1 has a directed interaction with receiver 2, 2 with 3, and 3 with 1.

2.2 ATTRIBUTE DATA

Relational data then, are the essential requirement of a network, and much of the work in constructing and analyzing animal social networks is concerned with these relations. Yet to make biological sense of a network it is essential that its structure be analyzed in terms of, among other things, the properties of the individual animals (see figure 1.4 and the associated text for a simple illustrative example). Such properties are referred to as attribute data. They could include genotypic or phenotypic variables such as an individual's sex, age, body size, coloration, and so on, or be used to characterize ecological traits of an individual such as preferred habitat use or home range, or even to express behavioral tendencies or the current motivational state of the individual. A simple example of an attribute data set is given in box 2.1. Though we have much less to say about the collection of attribute data than about relational

Box 2.1

Representing Attribute Data

Attribute data can be represented in a case-by-case table of the measured variables. Each row of the table represents an individual and the measured variables that are attributed to that individual. This form of data can be easily transferred into standard statistical packages, and statistical tests and descriptive analysis carried out. For example, in table 2.5 we could ask what the mean body length of the population is and whether it differs between males and females. Attribute data can be used alongside relational data when constructing the social network (chapter 3) to provide insight into factors contributing to network structure.

TABLE 2.5.
An example of a table containing attribute data for guppies, showing the sex and body size of the individuals.

Individual	Sex	Length (mm)
309	Female	34
203	Female	37
102	Male	23
223	Male	19
101	Female	29

Figure 2.3. Examples of natural variation in markings that can be used for individual recognition (reticulated giraffe, *Giraffa camelopardalis reticulata*; plains zebra, *Equus quagga*; manta ray, *Manta birostris*; killer whale, *Orcinus orca*).

data, we will return many times to the role of attribute data in making sense of the structure of animal social networks.

2.3 IDENTIFYING INDIVIDUAL ANIMALS

To construct social networks in which one node represents one animal, it is not enough just to observe interactions or associations between individuals. We also need to be able to identify those individuals repeatedly and correctly. Essentially, there are two ways that animals may be identified. They may be identified using natural marks, which has the advantage of being noninvasive, or using artificial tags or marks, which may increase the accuracy of identification and may also allow for automated data collection. In this section we will review some of the methods for identifying individuals and some of the problems that may be encountered with the various techniques.

Natural Markings

In many species individuals can be identified using natural markings that may be compared to reference photographs or drawings (see figure 2.3). In addition to using natural morphological markings, many animals may have acquired injuries and have scar damage or lameness that can be used to aid individual recognition, though such features may only be useful for a limited time span. One important consideration with using natural markings to identify individuals is the accuracy with which individuals can be re-identified on a second sighting (especially if a considerable time period has elapsed between observations). Clearly a high degree of misidentification is going to produce a meaningless network or, worse still, a network whose analysis might lead to erroneous inference about the study system. Such errors may be reduced by using computer-assisted photo identification to recognize individual animals in a population (e.g., Beekmans et al. 2005). A few of the different problems that may be encountered when using natural markings are discussed in box 2.2.

Box 2.2

Identifying Individual Reticulated Giraffe

Each study system will have its own issues that need to be taken into consideration when using natural markings on individuals. We will illustrate this by using a recent investigation of the reticulated giraffe (*Giraffa camelopardalis reticulata*) by Shorrocks and Croft (2006), who identified individuals using neck markings. Each of the yellow lines that reached the mane were categorized according to the angle they made with the mane. There were three categories of line: right-angled (R), acute-angled (A)—making an angle less than 90 degrees—and obtuse-angled (O)—making an angle between 90 and 180 degrees (see figure 2.4). Each animal is thus coded by a series of Rs, Os, and As, reading the yellow lines down from the head.

As there are several (typically 10) yellow lines on each side of a giraffe's neck, three types are enough to produce many more permutations of line-type along the neck than there are giraffes in a local population, yielding what is likely to be a unique recognition pattern for each animal. The 10 positions on one side of the neck yield $3^{10} = 59,049$ possible arrangements. If this isn't enough, viewing the other side of each neck as well gives a plentiful 3^{20} or 3,486,784,401 potential different combinations! (Of course, all this variation is only useful if we know that neck marks are truly individual, and not inherited, for example.)

It is important to evaluate the probability of making a mistake with such a technique. In this study there are two types of mistake, namely, identifying two individuals as being the same and misreading an individual already seen. The first issue can be resolved analytically. The second, we believe, can only realistically be solved empirically.

One way to address the problem of identifying two individuals as the same individual is randomly to sample a virtual population of neck codes, with the same frequency of R, A, and O as actually observed. In the sample of 100 giraffes investigated by Shorrocks and Croft, the approximate frequency of the three types of line were: R 52.6%, A 22.3%, and O 25.1%, with most necks having at least 10 lines. One thousand random samples of 10 lines (corresponding to one side of the neck) were taken from such a population and the same code occurred 28 times. With 20 lines (both sides of the neck) no code was observed twice. This means that there is a low chance (<2.8 in 100) of getting two giraffes with the same pattern if only one side of the neck is observed. With both sides observed this probability becomes very small ($p<0.001$). The actual probability of misidentification will in fact be less than these figures suggest because each individual has additional attributes that can be used for identification, such as age (e.g., adult, juvenile) and sex (male, female).

The probability of misreading an individual already seen is dependent on the accuracy with which observers can repeatedly code the same neck pattern. Observers will have an inherent probability of making a mistake that is not random and which we can only quantify through empirical investigation. To address this second issue Shorrocks and Croft took photographs of 9 individual giraffes. The observer in the field coded the neck patterns 5 times, with a 15-minute break between each scoring. Over 90 percent were identified correctly.

Marking and Tagging Individuals

In the absence of sufficient inter-individual variation in phenotypic markings, it may be necessary to mark individuals artificially (see Lane-Petter 1978; Twigg 1978; Martin and Bateson 2007). When using artificial marks on individuals, it is essential to consider whether the mark may affect behavior. If the mark itself influences the association patterns, then the resulting network may not be representative of the natural network. A classic example that illustrates this point comes from an investigation of zebra finches by Burley (1988). The finches were marked using colored leg bands, which affected the association patterns. Female finches preferred males wearing red leg bands over un-banded males. In contrast, males preferred females with black leg bands. Both males and females avoided members of the opposite sex wearing green or blue leg bands.

This example illustrates the importance of assessing the effects of a marking technique on associations or interactions before deploying the technique to

Figure 2.4. The three types of lines occurring in the neck pattern of reticulated giraffes: right-angled lines (R), acute-angled lines (A), and obtuse-angled lines (O).

investigate social structure in the laboratory or field. Of course, the effect of any marking system need not be particularly subtle. Marks or tags may influence the mobility or stress levels in animals, and thus may change their behavior.

2.4 DESIGNING A SAMPLING PROTOCOL

When developing a method to sample populations, it is essential that it is appropriate for the study system. The sampling protocol will depend on the population size and the ease with which individuals and interactions or associations between them can be identified. As is so often the case, the ease of observation may be very different for investigations conducted in the laboratory (or on captive animals) and those conducted in the field. A number of investigations have examined the structure of social networks in captive animal populations, particularly primates (Sade and Dow 1994), with a restricted number of individuals confined and habituated to humans, making it relatively easy to observe inter-individual interactions between all individuals in the population.

When the population is small, it may be possible to sample the population continuously and collect essentially complete information on social interactions. In the field it will often only be possible to construct a network via a series of samples or censuses. Then we need to consider the independence of consecutive sampling events. It has been shown, for example, that the time frame used to calculate association indices (see chapter 3) may have a great effect on the apparent structure of a population (Cross et al. 2004). As a general rule the time interval between sampling events should be greater than a typical mixing time for the animals; in other words, individuals need to have had the opportunity to exchange who they are interacting or associating with between consecutive samples for them to be considered independent. The time period over which this occurs will obviously depend on the species and, in the case of group-derived associations, how frequent fission and fusion events are. For example, guppy shoals disperse overnight (resulting in the breakdown of shoal composition) and reassemble every morning (Croft et al. 2003). Thus samples of group membership in this species made on consecutive days can reasonably be assumed to be independent (Croft, Krause, and James 2004a). However, if sampling occurs faster than the opportunities for individuals to exchange between groups, then significant false trends may be seen that are independent of actual individual association preferences (Cross, Lloyd-Smith, and Getz 2005).

One of the important questions regarding empirical research on networks is that of the sampling effort. How many times do we need to sample a population to get an adequate picture of the "real" network? As always in statistics, the larger the number of samples, the better. A good starting point for collecting social network data is to obtain an estimate of the size of the study population. This can sometimes be done directly if the animals are easy to count or if numbers are small. Otherwise we refer the reader to the numerous mark-recapture techniques mentioned in most ecology textbooks (e.g., Krebs 1998). Once we have information on the population size, we can decide what proportion of individuals can realistically be marked (or identified) and observed over a certain time period. Thus the proportion of individually identifiable animals out of the total population is a first good indicator of how likely we are to get anywhere near a reliable estimate of the population network.

Once we have started sampling the population, we can gain some idea of how well we are doing if we monitor some of the properties of the "growing" network or data set as it is being accumulated. For example, in an investigation of guppies we observed that nearly 80 percent of individuals initially marked and released were interconnected into one network component after just two sampling days (figure 2.5). We also looked at the time span over which interactions occurred and the number of times individuals were resampled (figure 2.5), in order to get some idea of whether our total sampling effort was likely to be enough to produce useful, statistically analyzable networks.

At first glance the issue of sampling populations for social network analysis might seem relatively straightforward; surely we should just follow the general principles of sampling ecological systems (Krebs 1998). So if we take a representative random sample of individuals, then we might expect the network constructed from the relations among them to be representative of the population as a whole. Unfortunately, the situation is slightly more complicated than this (Alba 1982). The problem arises because we are deriving relational data from our samples and, particularly in a fission–fusion system, the number of relations among members of the sample will be only a subset of all their relations, and they may not be representative of their social relations across the population as a whole. This problem will be reduced by repeatedly sampling the same populations of individuals through time, and the more times the population is sampled, the more likely we are to get a representative picture of a given individual's social network of relations.

So how much information do we lose when deriving relational data from sample data? Unfortunately it is difficult to give an authoritative answer to this question, though we can take some hints from the social sciences literature. For example, Burt (1983a) estimated that for human social networks, the amount of information lost is approximately $100-s$, where s is the sample size expressed as a percentage of the population. Thus if only 10 percent of the population is sampled, 90 percent of the relational data will be missing. However, such a calculation assumes that we are able to get complete data on the individuals sampled. While this may be true in humans (through questionnaires for example), in other animals this will be much more problematic because we can only assess who the animal is associating with at a given point in time, and cannot ask the individual about its past and intended future interactions, as in a questionnaire survey in humans. Thus we will ultimately only have partial data on inter-individual relationships, reducing the data even further. However, once again repeated sampling of a population can help to overcome this problem.

Given that most social network data will be incomplete, how will this affect the measured properties of the network? Some attention has been given to this topic in the social sciences (Holland and Leinhard 1973; Laumann, Marsden, and Prensky 1983; Bernard et al. 1984; Kossinets 2006). For example, Costenbader and Valente (2003) and Borgatti, Carley, and Krackhardt (2006) have looked at the effect of sampling on some important measures of natural structures (also see Lee, Kim, and Jeong 2006; Yoon et al. 2007; and Stumpf, Wuif, and May 2005). Others have focused on the so called "boundary specification problem" (Laumann, Marsden, and Prensky 1983). This refers to the task of

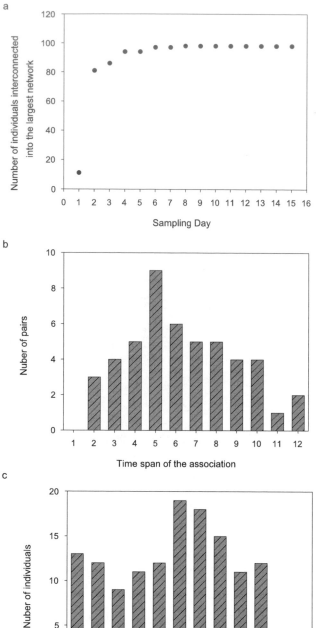

Figure 2.5 Measures of the resampling profile for a population of wild guppies resampled once per day for a total of 15 days by Croft et al. (2005). (a) The number of individuals that were socially interconnected, (b) the timespan (days) over which associations were observed, and (c) the resampling frequency of individuals.

deciding which individuals to include in the network and which to exclude because, for example, we do not believe that their interactions are well represented by our data. If we are conducting our network investigation on a closed system, such as a small population of captive primates (Sade and Dow 1994), then boundary effects will not be an issue because the population is isolated and all individuals and their interactions are likely to be equally observable. In investigations on wild populations there will be individuals on the periphery of the network that we may have sampled infrequently. Often as biologists we will see merit in filtering the network to reveal the core social structure (see the sections on filtering networks in chapter 3 and 5), defining our boundary based on some measure of association strength or frequency of observation. However, very little research has been conducted on the possible effects of removing individuals and edges on the measured properties of the animal social networks (see Wey et al. [2007] and, in a slightly different context, Lusseau, Whitehead, and Gero [2008]).

If our aim is merely to get an overall impression of the social structure of our population, then we may be able to live with an undefined or poorly defined boundary. We should expect that, provided our sample is big enough, average values of individual-based measures will be useful quantities. If, however, we wish to make a quantitative analysis of the exact social neighborhood of particular individuals, we need to be more careful. One avenue that might then be worth exploring is to adopt a form of sampling used in the social sciences known as snowballing (Goodman 1961).

Sampling Populations Using a Snowballing Technique

In a snowballing sampling technique we initially draw a sample of individuals from a population. These individuals comprise the "first order zone" (Wasserman and Faust 1994). The aim of the method is to get an accurate sample of the social connections of the individuals in this first order zone. To this end, we resample the population, collecting information on all the contacts of the members of the first order zone (Frank 1978; Frank 1979). Any individuals that are directly connected to the first order zone but are not themselves part of it are defined to be in the "second order zone." Newly observed individuals not directly connected to those in the first order zone are defined to be in the third order zone, and so on. The sampling continues until there is sufficient information on the direct and indirect network relations of individuals in the first order zone to be confident of the social structure of individuals within that zone. After all this, we still need to decide for how long we should sample the population and for how many zones. The references quoted above offer some guidance here. Of course, if we are going to use this method to comment on the population properties, then we must assume that the segment of the network sampled is representative of all other segments in the network.

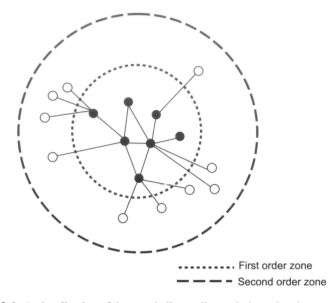

First order zone
Second order zone

Figure 2.6. A visualization of the snowball sampling technique showing two sampling zones. The first-order zone contains the 7 individuals (black nodes) initially sampled in the population. The second-order zone contains individuals that have a direct network contact to individuals in the first-order zone. The circles indicate the boundaries of the zones.

While this discussion may provide some answers to the issues of sampling populations to establish social networks there is a need for future work to investigate these issues quantitatively. For example, with computer processing power now so cheap it should be possible to simulate large "virtual populations" of animals, with realistic patterns of interaction, and to subject these populations to various sampling protocols. This would make it possible to compare sample networks with population networks built on complete information, and provide quantitative information on the key questions such as: For how long do we need to sample a population? What proportion of the population needs to be sampled? How can we scale up information on a sub-network to the population as a whole? Such an approach is eagerly anticipated.

2.5 MANAGING AND PROCESSING NETWORK DATA

Social network analysis often involves dealing with large data sets, and it is essential that the data be organized in an appropriate way for easy conversion to an association matrix. Transforming raw "field" data into a matrix can be

TABLE 2.6.
Examples of data entry files in a linear mode (a) and group mode (b). The data sets
are subsets of the examples provided with SOCPROG. The first column in each table
represents the date and time the observations were made. In the linear mode table (a) the
group that an individual was observed in is coded in the column labeled "group" and
the identity of the individual is in the column labeled "ID". For example, individuals
A1 and I9 were both seen together in group 1 on 12/9/89 at 9 a.m. In contrast, in the
group mode table (b) all individuals within a group are placed within a cell in the group
column. For example, on 12/9/89 at 9:49 individuals 8, C11, A13, and 20 were all seen
together in the same group.

(a) *Date*	*Group*	*ID*	(b) *Date*	*Group*
12/9/89 9:00	1	A1	12/9/89 9:49	8 C11 A13 20
12/9/89 9:00	1	I9	12/9/89 14:54	A1 9 A14 A15
12/9/89 9:00	1	N14	12/9/89 15:41	4 7 A12 A17 A19
12/9/89 9:00	1	O15	12/10/89 9:11	4 7 A12 A17 A19 20
12/9/89 12:00	2	H8	12/10/89 9:41	2 A10 A18
12/9/89 12:00	2	K11	12/10/89 10:09	D3 5 6 A16
12/9/89 12:00	2	M13	12/11/89 10:35	2 A10 A18
12/9/89 12:00	2	T20	12/11/89 11:03	4 7 A12 A17 A19 20
12/9/89 15:00	3	D4	12/11/89 14:32	5 6 A16
12/9/89 15:00	3	G7	12/11/89 17:40	A1 9 A14 A15 8
12/9/89 15:00	3	L12		
12/9/89 15:00	3	Q17		
12/9/89 15:00	3	S19		

a time-consuming process if done manually. Happily, several programs are
available that will make this conversion. For the purpose of this book we will
focus on two programs that we have found to be particularly useful for processing and managing network data: SOCPROG and UCINET (see box 1.1). We
will use these programs to illustrate the general features of managing network
data. The principles for working with other packages will be very similar, but
the specific details may vary slightly.

In our own research we have found SOCPROG very useful for converting
information on group composition into an association matrix that can then be
imported into a social networks package. SOCPROG is a very adaptable package
and data can be entered in two forms, in a "linear mode" or in a "group mode"
(see table 2.6 for examples). SOCPROG comes with a very comprehensive and
user-friendly help manual, and we refer the reader to this for further details of
constructing association matrices from data files.

Alternatively, data can be imported directly into a network analysis package
such as UCINET. This allows you to load data in several formats, some of which
we discuss below. Once in the program, the data is stored and manipulated via
two related UCINET files. The first file will contain the header information and

TABLE 2.7.
Example of a DL file constructed for a population of
8 individuals labeled A–H.

dl n 8 format = fullmatrix
labels: A,B,C,D,E,F,G,H
data:

0	16	18	3	19	6	8	7
16	0	16	3	19	8	8	10
18	16	0	2	29	7	9	7
3	3	2	0	3	8	1	2
19	19	29	3	0	6	8	8
6	8	7	8	6	0	6	3
8	8	9	1	8	6	0	5
7	10	7	2	8	3	5	0

the identities of the individuals in the network, and is saved as filename.##h. The second file will contain the relational data and is saved as filename.##d.

So, how do you get data sets into UCINET? One laborious method is to enter your data manually into UCINET using the spreadsheet editor: in UCINET go to *data > spreadsheet editor*. The spreadsheet editor can also be used to manipulate saved files.

Various other format of data can be imported into UCINET, including so-called incidence matrices, and it is worth exploring some of these before deciding how to record your relational data. Perhaps the most flexible way to import data into UCINET is to use a "DL" file format. The "DL" stands for "data language." It is a language that codes information to be imported into UCINET with the minimum amount of data entry. Most usefully, there are many different formats that can be used to code data in DL files, reducing the likelihood of your having to reenter your data before analysis. For full details of the DL code we refer you to the very good reference guide that comes with UCINET. However, to get you started we will illustrate a simple example.

Table 2.7 illustrates some important features of a DL file. The file starts with "dl" so that it can be identified by UCINET. "*n*" is the number of individuals in the sample, "format" specifies the form that the data is entered in, and "format = fullmatrix" indicates the data are in the form of an association matrix, with all elements to be entered. Since there are many different formats that can be used to code data in DL files, the choice of which to use will depend on the form of your data. The data is then given in the subsequent lines of the file with one row of the association matrix per line in this particular format. Note that the values on the diagonal are coded as "0." If you are going to use UCINET to explore your network data, it is worth spending a little time becoming familiar with DL files, as they will allow you to enter your data into UCINET with minimal effort.

The entry of attribute data is more straightforward. It can be achieved in UCINET via, among other things, a "vna" file, which is a simple text file with a straightforward format comprising lists of animals and their attributes in subsequent columns. We will make use of both relational and attribute data files in the next chapter and beyond.

3

Visual Exploration

So, now that you've enjoyed (or suffered) the collection of your data and used it to compile a social network of some sort, it is time to start thinking about what that network can tell you about your study system. As humans we are exceptionally good at visual pattern detection, and in any data analysis, a sensible first step is to visualize what you have. Just as we may plot a histogram or scatter graph for more conventional data, we can represent the social associations or interactions that we have measured as a form of graph that, as we have already seen in chapter 1, represents each animal as a node and each pair-wise relation as an edge or line between two nodes. We should reiterate at this point that the data matrices introduced in chapter 2 contain exactly the same information as (and sometimes more than) the graph, so it doesn't really matter whether we describe our data as a "network" or a "graph" (though some mathematicians, in particular, seem loathe to use the word "network," for reasons that aren't entirely clear to us). As we will see in later chapters, much of the quantitative analysis of networks is most conveniently calculated using the matrix representation of social connections. But first we should see how much there is to learn from a graphical representation. Many of the issues we will raise in this chapter are central to the remainder of this book; it is our hope that the simple qualitative network descriptions covered here will encourage you to want to ask quantitative questions of your data, and help to decide exactly what those questions should be. We will turn to more quantitative matters in chapter 4.

In this chapter, then, we will introduce you to various standard ways in which we may visualize social networks, and demonstrate how simple manipulations of a graph may be used to begin to reveal information about the social structure of a population of animals. There are a number of steps that might typically be taken. The first (section 3.1) is simply to draw the network, though as we shall see, some drawings are more pleasing to the eye (and more useful to the brain) than others. Rather clever algorithms such as "spring embedding" can be used to arrange the layout of the network based on the closeness of interactions between individuals, and so reveal interesting structural features. Network edges can be drawn with an arrow to represent the directionality of social interactions, and with a thickness representing their weight (see chapter 2 for an explanation of these terms). Where available, we can incorporate node (animal) attribute data into our visualizations, and investigate whether these attributes contribute to the observed network structure.

The simplest way to achieve all this, and nearly everything else covered in this chapter, is to use one of the many computer packages available. Visualization programs have the advantage that they control for the anthropomorphic influence due to preconceptions or misconceptions of how we think the network should look. Furthermore, for networks that contain more than just a few individuals, computer-drawing packages allow the network to be visualized in seconds, whereas it may take hours or days to draw by hand. As an additional bonus, many drawing packages offer some simple analysis and manipulation tools, as we shall see. Which of the programs you opt for is largely a matter of personal taste (see Huisman and van Duijn [2005] for a review of network analysis and visualization packages). Some can produce only relatively simple two-dimensional monochrome visualizations; others are more sophisticated. To illustrate the principles of network visualization, we will focus on one package, NETDRAW (Borgatti 2002). NETDRAW can be used as a stand-alone program or as part of the UCINET network analysis package (see box 1.1).

With the network drawn, we can observe some of its core structural features. One of the first things we will notice (section 3.2) is whether or not the network is composed of a single "component" with all individuals connected, directly or indirectly, to all others in the population. We may also want to focus our attention on the relative position of individuals within their social network (section 3.3), identifying those that appear to be key in connecting different regions of the network or constructing and comparing the network neighborhood of one or more chosen individuals.

A useful feature of NETDRAW (and similar packages) is that it gives us a quick way to explore how robust our network is to filtering the data by including only those interactions that have occurred a certain number of times (section 3.4). We have deliberately chosen one of the illustrative examples for this chapter to be a case where such filtering reveals that there is insufficient data to draw too many conclusions, and another where it reveals one of the interesting biological questions to be asked in that case. We will also discuss the merits of using "association indices," rather than a simple count of associations, to filter the edges. Finally, we might want to consider removing individuals from our network that have not been sampled frequently enough to have confidence that we have a true depiction of their social interactions. Again, with a little forethought when collating our data, visualization packages can be used to see where such removals should occur in the network.

In our experience, a visual exploration of your data as a network of nodes and edges is very likely to reveal some patterns you expected, and some you may not have. Seeing the intertwined relationships between the animals you have so painstakingly studied flashed up on a computer screen is an exciting and rewarding experience. After a little juggling, you may quickly form hypotheses to explain the observed social structure. We strongly suggest that you play with the different methods available, as each may reveal different patterns in the data. Of course, there is a danger that we see patterns where they do not really exist. It is therefore essential not to rely on visual analysis alone but to undertake quantitative analysis of the network to test the significance of the observed network structures. We return to these issues in the remaining chapters of the book.

3.1 DRAWING SOCIAL NETWORKS IN NETDRAW

Let us get started. Take your network and load it into a visualization program such as NETDRAW (see box 3.1). Unless you have very few animals in your study system, the network you see is likely to be a bit of a mess, with nodes (red circles in NETDRAW), edges (black lines, all with arrowheads next to the nodes), and text labels all getting in the way of each other. Don't despair! Unless you have several hundred animals or more, things can easily be made better. Node labels and arrows (which are unnecessary for an undirected network) can be toggled on and off with buttons at the top of the NETDRAW screen. The circles can all be made smaller or larger, and the lines thinner or thicker using functions listed under the "Properties" menu. The most helpful capability, however, is to be able to manipulate the layout of the nodes on the screen so that important structural features can begin to be seen.

Some manipulation of the layout can be achieved by hand. In NETDRAW, holding down the left mouse button over a node enables you to drag it anywhere you like on the screen, with all the edges to and from that node remaining intact. It would be much more useful, of course, to have some automated means of moving the nodes around so as to reveal network structure. Several of these are listed under the "Layout" menu in NETDRAW. By far the most useful layout scheme, we have found, is "spring embedding." Though it sounds like something that might be more at home in a gardening book, spring embedding is in fact a neat trick to achieve what we need—a layout of the network with densely connected nodes clustered together and nodes with few connections placed around the edge. Spring embedding achieves this by treating the network as a collection of masses (the nodes) connected by springs (the edges) and letting the springs pull the masses around until they reach a state of equilibrium. Groups of well-connected individuals tend to be bunched together in the resulting visualization (see figure 3.1). It should be noted that different starting positions will yield different spring-embedding visualizations of a network. One way to ensure that there is no bias in the visualization is to reset the graph to random positions. In order to find the optimal position, it is essential to give the spring-embedding algorithm sufficient time for the "masses" and "springs" to settle their positions. We have used spring embedding to visualize the network

a

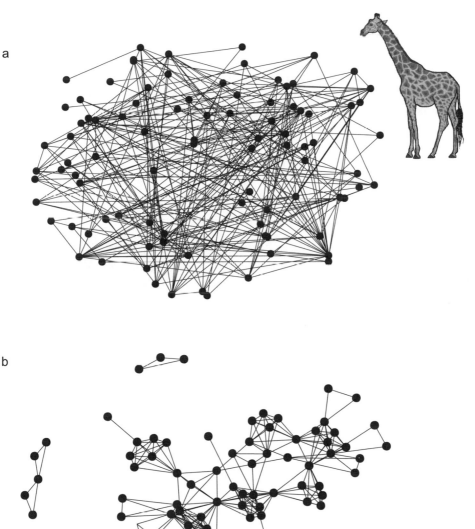

b

Figure 3.1. The social network of a population of reticulated giraffes in Kenya using (a) random and (b) spring embedding layout options in NETDRAW.

shown in figure 3.1, which is the first of our illustrative examples for this chapter. It is a social network for a population of giraffes (*G. camelopardalis reticulata*), drawn with two different layouts: (a) random and (b) spring embedding. The second of these clearly exhibits more structure than the first, and is much more inviting of further investigation. Nearly all of the networks in this book have been drawn using spring embedding, plus the occasional movement of some nodes by hand to tidy things up, as the layout scheme.

Box 3.1

Getting Started in NETDRAW

NETDRAW can be used as a stand-alone package or, more usefully, as an option from within UCINET. The NETDRAW menus make it very easy to load files of various formats. For example, if you have already prepared an association matrix and saved it as the UCINET file NET.##h, simply opening this file from NETDRAW will draw it to the screen. There are other ways of importing data as association matrices or DL files into NETDRAW (see the help files that come with UCINET, and chapter 2 for further information). The layout of the nodes and edges can be controlled via the options in the *layout* menu. A sensible start is to set the layout of the network to random. We might then use the spring-embedding function (also within the *layout* menu) to produce a more useful and revealing visualization of the network. The number of iterations run by the spring-embedding algorithm can be adjusted in a control box. The default value is 100.

Weighted and Directed Edges

If the edges in our network are either weighted or directed or both (see chapter 2) this should be reflected in its visualization. We can illustrate this approach through an example. In an investigation of Rhesus monkeys (*Macaca mulatta*), Sade (1972) collected data on the patterns of grooming behavior by recording "who groomed whom." The network for these data is shown in figure 3.2. The edges are both directed (in that not all grooming events are reciprocated) and weighted (multiple grooming events were recorded). The directionality of the interaction will automatically be represented in the form of the arrowheads (which may be switched on and off in NETDRAW). Thus in figure 3.2a we may easily see that individual 12 groomed 1, but 1 did not groom 12, but we cannot tell (at least from the visualization) how much grooming was observed

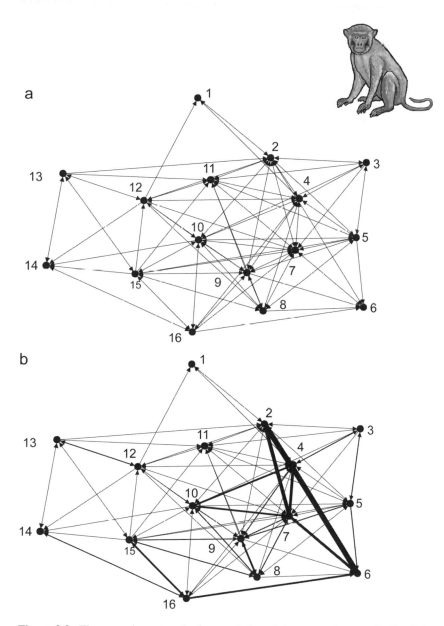

Figure 3.2. The grooming network of a population of rhesus monkeys studied by Sade (1972). In (a), the directionality of interactions is illustrated by the arrows connecting individuals. The weighting of interactions is represented by the thickness of the lines connecting individuals in (b), where a thicker line indicates a stronger (in this case, more frequent) interaction. Individuals have been relabeled from the original for ease of interpretation.

between a pair of animals. This can be achieved by using the thickness of lines to represent the strength of an edge. In NETDRAW, the function *properties > lines > size > tie strength* does the job, the result of which is shown in figure 3.2b (with thicker lines having higher "tie strengths," which in this case simply means the number of grooming events). We note in passing that an interpretation of the line thickness is straightforward for undirected networks, but a single parameter cannot convey all the information about multiple two-way directed interactions; it would take more to distinguish eight interactions from A to B and two from B to A from five interactions in either direction, for example.

A quick visual inspection of figure 3.2b shows that not all individuals engage in equal grooming activities. Indeed, there appear to be a few key individuals that have a high tendency to groom and be groomed (individuals 2 and 7, for example). Furthermore, some of the animals that are strongly connected (such as 2 and 6) have not been placed close to each other by the spring-embedding algorithm, which should be a reminder not to read too much into the layout of your nodes.

As a final aside in this short section, it should be stressed that the grooming networks in figure 3.2 are rather small, containing 16 individuals. This is a convenient size to plot as a graph in a book, but it perhaps does not do justice to the potential visualization and analytic power of a networks approach. It is quite likely in this case that most of the features that we noticed via visualization were readily apparent in the data as they were originally recorded. The real power of network analysis comes to the fore when there are too many nodes to be able to tell immediately from the data set what patterns are there and whether they involve small numbers of individuals or a larger fraction of the population. The next section gives us a means to help us start to look for these patterns.

Nodes with Attributes

One of the real keys to extracting quantitative information from an animal social network is to relate attributes (such as phenotypes) of individuals to their position or environment in the network, in order to probe the role of individual phenotypes in structuring social interactions. Thus a useful first step is to be able to codify individual attributes into the graphical visualization of a network. The most obvious attributes to consider are physical features such as sex and size. It might be even more revealing to measure and record behavioral attributes of individuals if they display consistent behavioral phenotypes such as boldness and shyness (Sih et al. 2004). Were we armed with information on, among other things, the sex, age, and dominance of individual rhesus monkeys, for example, this might lead to testable hypotheses regarding the grooming interactions in the population represented by the networks in figure 3.2. Of course, when using more than one attribute, we have to be careful about their possible interdependence. If larger individuals in a population are also females, it may be difficult to determine whether it is body size or sex that is responsible for driving network structure. This is a problem for later; for now we can merely use node attributes to aid our visual exploration of the data.

The NETDRAW program enables us to differentiate between nodes by varying their color, shape, and size (see box 3.2). Shape, and to a lesser extent color, may best be reserved for categorical attributes such as sex (male and

---

Box 3.2

Incorporating Attribute Data into NETDRAW

As we mentioned at the end of chapter 2, node attributes can easily be read into the NETDRAW program using a "vna file." The file has a very simple format; the first line tells NETDRAW that this is a vna file. The second line labels entries in the subsequent lines. The rest of the file consists of one node per line, with identities followed by values of the attributes associated with the node. So if we have five animals, labeled A–E, and wish to include their sex and age as node attributes, the vna file is simply:

*Node data		
ID	sex	age
C	1	1
A	1	6
B	1	5
E	2	7
D	1	7

Note that the lines containing the attributes can appear in any order. To import a vna file, use *file > open > vna text file > attributes*. NETDRAW will show the file in a "Node Attribute Editor," which can then be closed. Now the attribute data can be used as part of the visualization process.

We have found that the most appealing way to represent categorical attributes is to use different colors or shapes for nodes in different categories. In NETDRAW this is achieved by selecting *properties > nodes > color* (or *shape*) *> attribute based*. When the attribute data is continuous, it is more useful to manipulate the size of the node according to the attribute value. This can be achieved simply in NETDRAW by selecting *properties > nodes > size > attribute based*. The program will offer you a range of node sizes to represent the range of attributes.

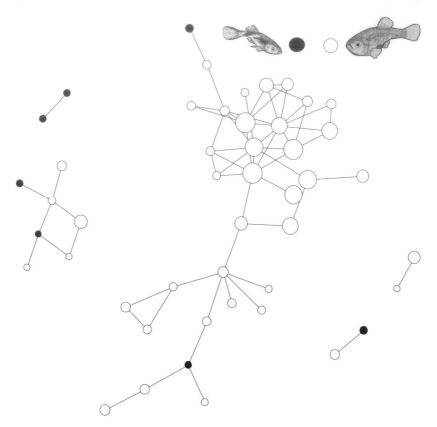

Figure 3.3. The social network of a population of Trinidadian guppies, illustrating the use of node shading and size to represent animal attributes. In this network, nodes representing female guppies are open circles, and males filled circles. The size of each node is a reflection of the body length of the fish; in this case the scale of the node sizes is arbitrary to exaggerate the differences between large and small fish.

female) or species, with each individual falling into one attribute. Node size may also be used to distinguish categories, but is more useful if the attribute is measured on a continuous scale; obvious examples would be body length, age, and weight. Should you wish to represent more than one continuous attribute in your visualization, you could consider converting one of them to a discrete attribute by binning. As an example, the size of individuals might be placed into either a small, medium, or large class, and three colors or shapes used to represent the nodes falling into each bin.

   Figure 3.3 shows a social network of a wild population of Trinidadian guppies (*P. reticulata*), sampled by Croft et al. (2006). Two of the phenotypes known to play a role in the social organization of this species, at least at the level of the shoal, are sex and body length, with the two covarying to the ex-tent that most large guppies are female (Magurran 2005). Thus it is natural to represent the sex and body length of individuals in a visualization of a guppy network. In figure 3.3, node shading is used to distinguish the sexes and node size the body length. A further look at this visualization reveals some poten-tially interesting features of the network. If we look carefully at the top of the network, we can see a cluster of large open circles (large females), suggesting that these female guppies may form the "core" of the network, a point that we will return to later in the chapter.

3.2 NETWORK COMPONENTS

Now that we have drawn our network to best advantage, we are ready to ana-lyze it a little further. We turn first to some useful preliminary analysis that can be performed from within a visualization program such as NETDRAW. One of the first things that you may notice when you first draw a social network for your system is whether or not it consists of a single component. A network component is a group of nodes (individual animals) that are interconnected, but with no connections to the rest of the network. Some networks contain only one component, others contain more. We might expect there to be many components if there are very few connections between animals; conversely a closed population of freely intermingling animals observed over a long period might be expected all to be connected in one component. Whatever the case, if we take our data at face value, the number and size of the components are likely to be of interest. Understanding the number of sub-network components might well give a first indication of how socially fragmented the population is. If individuals (or components) are isolated from a main large component, the first question to ask might be whether there is anything that identifies these network stragglers in terms of their measured attributes (see previous section).

In many cases, where the number of nodes is not too great and spring em-bedding does a good job of laying out the graph, it is easy to spot the compo-nents in a visualization. This is not always the case, though, so a useful feature of NETDRAW is the *analysis > components* function, which enables you to use color (or shape) to identify components quickly and easily. We have done this for the giraffe network of figure 3.1; the result is shown in figure 3.4. The net-work is seen to be composed of five different components, each represented by a different node shape. (In fact there are ten other components not shown in the figure. These are "isolates," or nodes that have no connection to any other in the network. They each form a component of size 1. They were not drawn in figures 3.1 and 3.4 to avoid cluttering up the graphs. However, they should not be ignored in a population analysis of, for example, the average number of network neighbors.)

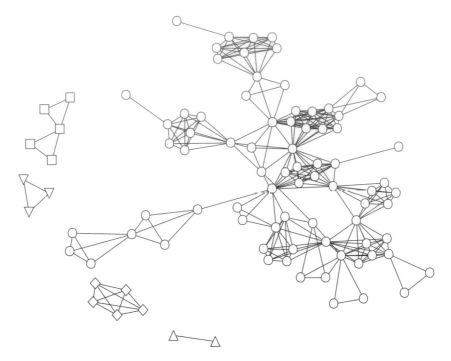

Figure 3.4. The social network of a population of Kenyan giraffes. Animals in the same network component are represented by the same symbol. This is the same network as in figure 3.1, and is again laid out using spring embedding. Note, though, that you have to look quite hard to see for sure that it is the same network, as much of the layout is arbitrary.

3.3 FOCUSING ON INDIVIDUALS

One of the main thrusts in (human) social network analysis has been to focus on the role of particular individuals in the network, in an effort to identify key players in the population (see, for example, Wasserman and Faust [1994] or Scott [2000]). Many measures have been developed, some of which we will meet in later chapters, to quantify the "centrality" or "prestige" of individual agents in a network. Such an approach has also been used in the study of relatively small networks of nonhuman primates (Sade et al. 1988; Sade 1989), but less so in more recent applications of network theory to other animal social systems, and perhaps with good reason. Any error in identification of an individual animal (see chapter 2) will, of course, give rise to an error in the network. One way in which this can occur is if two individuals are mistaken to be the same individual on one or more occasions. The effect of these errors

Figure 3.5. The social network of Kenyan giraffes from figure 3.4, illustrating blocks (black nodes) and cut-points (white nodes).

might be reduced by filtering the network edges to keep only the stronger interactions between pairs of animals (see later in this chapter), but even so, we should be more trusting of statistical measures, taken over all individuals (see chapter 4), than we are of focusing on any one individual, unless we are extremely confident about our ability to identify each animal perfectly. If we are confident, then some of the visualization tricks in this section, among others, may be of great benefit.

Blocks and Cut-Points

One way we may explore the role of individuals in social networks is to investigate individuals that occupy "cut-points" in the network. Cut-points occur when two sub-networks would be separate components were it not for connections through a single individual, whose node is identified as a cut-point (figure 3.5). The parts of the network connected by cut-points are referred to as blocks in this context. We could envisage a situation where these individuals are key players for the transmission of information or disease between the network blocks.

More generally, the blocks may be thought of as groups of interconnected animals with rather weak connections to the rest of the network. The presence of such blocks might be suggesting that there is a level of structure in the social networks somewhere between that of the pair and the population. In chapter 6 we will consider in more detail some interesting recent methods from network theory aimed at finding groups of well-connected nodes in complex networks. The methods are far less restrictive than the block and cut-point construction in their definitions of what counts as a meaningful group, but they are concerned with similar questions.

Ego-Centric Networks

As its name suggests, an ego-centric network is one centered on one individual node. In fact, the ego–centric network for a given node is just the sub-network containing only the node itself and those connected to it via an edge. (In other contexts this is described as the neighborhood of the node in question.) We can demonstrate the potential usefulness of an ego-centric network through an example. In the grooming network of rhesus monkeys studied by Sade (1972) (see figure 3.2), we suggested earlier that individual 7 may be particularly important for the grooming network in the population. This observation is strengthened by visualization of the ego-centric network for this individual (figure 3.6a), which shows that it is directly connected to twelve of a total of fifteen available conspecifics in its grooming ego-centric network. It is also worth noting that, in this case at least, the spring-embedding process that was used to lay out the overall network (figure 3.2) placed individual 7 close to the center of the network. This, along with the size of its ego-centric network, may be suggesting that monkey 7 played a pivotal role in grooming in the population. In contrast, if we look at the ego-centric network for individual 1, we can see that it is only connected to three others in its ego-centric grooming network (figure 3.6b), and appears to have a much less important role. Of course, such conclusions can only be drawn safely if, as we discussed above, we have confidence in our sampling protocol, as error in individual identification may greatly affect the network at this level.

3.4 FILTERING NETWORK DATA

The presence of separate components or blocks, or variation in the size and structure of ego-centric networks, indicate the presence of heterogeneity in a social network. These heterogeneities are the very thing of interest in animal social network analysis: we want to probe the extent to which there is structure in the network of connections within a population. As well as these heterogeneities, there are likely to variations in the number (or intensity) of interactions between pairs of animals and, especially in wild populations, variations in the number of times each individual is observed. In chapter 5 we will discuss the

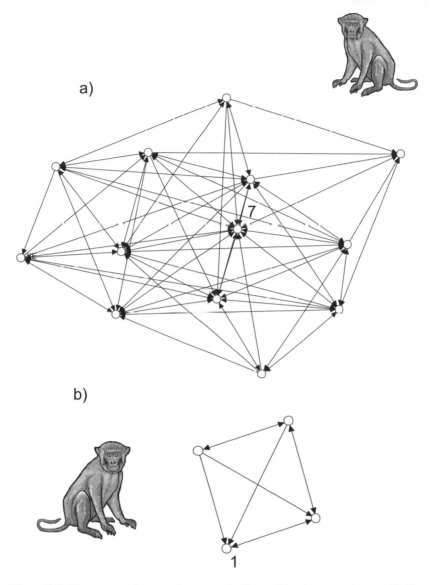

a)

b)

Figure 3.6. The ego-centric grooming network of two of the rhesus monkeys studied by Sade (1972): (a) individual 7 and (b) individual 1. See figure 3.2 for the full network.

importance of these issues when trying to establish the statistical significance of network-derived results. For now, we merely note that we should, as part of our preliminary visual exploration of a network, have a look at the effect of variable observation frequency and strength of relations on the network.

There are two ways in which we might examine these heterogeneities in our data, both of which involve filtering the network in some way. First, we may remove individuals that we have sampled very infrequently from the data set, assuming that we have less confidence that we have a true depiction of their social interactions. Second, we may remove weak or infrequent network relations (edges) from the network, so revealing the "stable core" of the network. The two methods we will consider are both based on the count of associations within groups. Edges can be filtered from the network simply based on the number of times individuals co-occurred in the same group. Later in the chapter we will consider the use of association indices to correct for certain biases in edge weights that might be a by-product of our sampling regime.

Before we go on, we feel it is only right to say that the issue of filtering in networks is one that seems to be subject to a certain amount of arbitrariness. Many studies are conducted on filtered networks, and there are definite benefits to this approach. Filtering social networks will help us to think about hypotheses that may explain network structure and to examine the robustness of emerging patterns. It may also highlight patterns in the network that we have predicted a priori, instilling further confidence in the network approach. In practice, though, most authors do not report that they have explored in any depth the effect of their filtering. Many have applied some sort of filter to their "raw" network, with a biological or sampling-related justification for doing so, then analyzed only the filtered network, treating the edges as binary (weight 0 or 1) and ignoring the possible effect of the animals or relations that have been omitted. Others just take the network as it is, but do not address whether their network might be subject to errors or sampling bias. We return to these issues in more detail in chapter 5.

Filtering Individuals

In investigations on wild populations, no matter how much effort we put into our sampling strategy, our data set is likely to be sparse, and some animals may be observed very infrequently. This may be because the individual has moved outside the study area or because of low recapture rates. The problem then arises that we may have observed some individuals too infrequently to be confident that we have adequately represented their social ties. Our first mode of filtering is to remove such individuals from the network. In an investigation on dolphins (*T. truncatus*), Lusseau (2003) only included individuals that survived the first twelve months of a six-year study, so that enough information was available to analyze their association preferences. When observations are conducted over substantially shorter time periods, we may want to consider thresholds set over a period of weeks or days.

As an alternative to a time threshold we could set an observation threshold. For example, we may decide that only those animals that were observed a minimum number of times are included. Wolf et al. (2007) were interested in the core social network of an island population of Galápagos sea lions (*Zalophus wollebaeki*), so they removed individuals that were seen fewer than ten times in surveys over a four-month period, as these were likely to be only casual users of the island. In a survey of the published literature at that time, Bejder, Fletcher, and Bräger (1998) found that authors often applied an observation threshold, which was typically set somewhere between two and six sightings.

Within the NETDRAW program, the easiest way to explore the effect of removing nodes according to criteria such as these is to include observation frequencies as attributes. So for example, to filter animals by the number of times they were observed, enter those numbers as an attribute, then check only certain values within the "Nodes" box to be included in the graph. Note that any graph you produce can be saved as a UCINET file, and subsequently subjected to the type of quantitative analysis you will see in chapter 4 and beyond, thus making it possible to perform a more systematic evaluation of the effect of filtering.

Filtering Edges

Having decided which animals to include in our network, we may now want to explore how the heterogeneity in the strength of network interactions or associations influences the observed network structure. A certain proportion of our observed pair-wise relations may have been due to random or "chance" events. (This is more likely to be a concern if our relational data are associations built on the gambit of the group, as we will discuss more in later chapters.) It is entirely possible that some of these chance events are rather important in holding a network together, and we should always be aware of the possible "importance of weak ties" (Granovetter 1974): it can take very few infected individuals carrying a virus to connect the whole world in a pandemic. However, especially given the dangers of over-analyzing network features that could be the result of misidentification, we might be better advised to focus on the relations that are more likely to represent nonrandom social preferences or interactions. Provided our edges are weighted (see chapter 2), we can start to identify the nonrandom elements of the network by filtering the network so that only those interactions or associations whose weight is above a certain threshold remain. As we increase this threshold, "core" components of the network may emerge. A simple method to filter edges in a network derived from group-based associations via the association strength (the number of times a pair of animals was observed in the same group). In NETDRAW, you can use the box titled "Rels" to control the level of an association-strength filter. It is possible to increase the association strength at increments of one and watch the associations (edges) that fall below this threshold disappear, revealing higher-order structure in the network. Again, any network on a NETDRAW screen (including filtered networks) can be saved as a binary network for subsequent quantitative analysis—in other words, to a certain extent, NETDRAW will do the filtering for you.

So what is the effect of all this in practice? We will look at two examples. The first is meant to offer a salutary lesson. The second shows that the extra information contained in the strength of the edges (beyond the fact that they are either present or absent) may be the key to unlocking some useful biology.

In our first example, we look again at the social network of giraffes shown in various guises in figures 3.1, 3.4, and 3.5. These networks look very promising for further analysis in that they are clearly very structured, with strongly interconnected regions and more sparse sub-networks, many cut-points, and so on. However, if we filter so that all connections occurring once are removed, only two associations remain! This should ring alarm bells. In fact, this is an example where the "gambit of the group" (see chapter 2) has not paid off. In constructing this network, all co-members of each group of giraffes encountered are assumed to be connected to each other. Most animals were observed only once, but just enough were seen again to join most of the population into a network with one large component. Thus, at least with this much data, there really isn't enough of a network to be analyzed. True social affiliations are only likely to be observed once each animal has been observed many more times. Defining all individuals within a group to have a direct network connection may lead to misleading network structures at low-association strength that simply represent the composition of groups at given sampling events. Thus in investigations that define associations based on the gambit of the group, it may be a good idea always to filter out edges between pairs of animals observed together only once. To have some confidence that we are observing meaningful social structure, it would be helpful to have a rule of thumb by which to set an edge-filter threshold. Measures such as the mean or median value of co-occurrence are obvious candidates; we return to the issue of how to set a filtering threshold in chapter 5.

Now we turn to a second example. In an investigation on guppies we marked all individual fish in a wild population using visual tags and recorded their social structure by observing who was in a group with whom over a seven-day period (Croft, Krause, and James 2004a). In addition to collecting data on the social interactions, we collected attribute data, including the sex of individuals. We incorporated attribute data and represented the sexes as different node colors in the network. If we look at unfiltered data in figure 3.7a, we see that all but two males can be connected into the same network component. The remaining graphs in figure 3.7 show the same network data filtered to include only pairs of fish that occurred twice or more together in a shoal (the median value) (3.7b) and three or more times (3.7c).

In this case the edge filtering reveals some interesting features of the population. First, and perhaps not surprisingly, the number of individuals in the

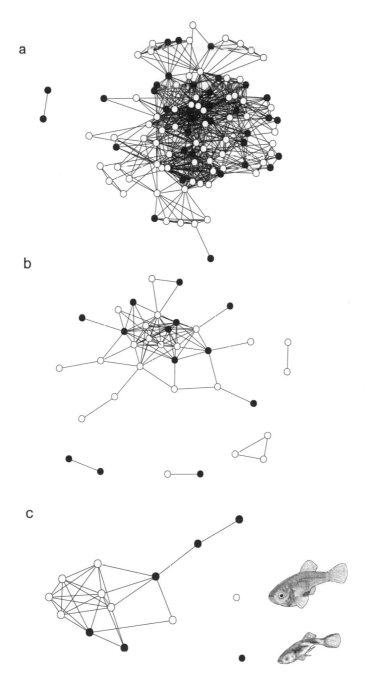

Figure 3.7. The social network for a population of guppies sampled by Croft, Krause, and James (2004a) (males = ●, females = O) drawn using spring embedding. Sub-networks are shown for different association strengths in which edges are only displayed between two fish if they were caught in the same shoal (a) at least once, (b) at least twice, and (c) at least three times.

largest network component decreases as we filter the network at higher association strengths. Second, a greater proportion of males drop out of the network as we increase the association strength, suggesting that it is females, and not males, that form the core social unit of the population. However, to investigate this statistically we would have to do more, and run a series of randomization tests to determine whether the observed female-female interactions occur more frequently than expected by chance (see chapter 5).

Association Indices

As an alternative to filtering the network based on association strength (a simple count of the number of times animals co-occurred in the same group), we could use an association index to identify the relative strength of interactions in the network, and then filter the edges based on the association index. Indices for calculating the frequency with which two individuals associate have been around for some time (e.g., Fager 1957), and they have been applied to a range of problems. Here we will focus on indices that have generally been used on fission–fusion societies where social associations are defined based on group composition. Other indices focus on associations defined via space use (see Wilkinson [1985] for an example). There are a number of advantages to using association indices over simply scoring the number of times two individuals associate. First, association indices can be used to correct potential sampling biases that have occurred during data collection. For instance, a bias may occur in investigations on cetaceans that use photo identification (Bejder, Fletcher, and Bräger 1998). In such studies only a small proportion of a given group may be photographed on any one encounter. As a result there may be a bias towards observing individuals in separate groups because not all individuals within a group are identified in a given sampling event. Second, where applied correctly (Cairns and Schwager 1987), association indices may allow between-study comparisons to be made of the strength of associations between individuals.

Various indices have been developed to compensate for different sampling biases. An outline of some of those that are relevant to group-derived data is given in box 3.3. More detailed discussion can be found in the papers by Cairns and Schwager (1987), Bejder, Fletcher, and Bräger (1998), Whitehead (1999), and Whitehead and Dufault (1999).

When using association indices we again need to decide what threshold value we should use to accept or reject interactions as edges in our network. Attempts have been made to assign significance values to dyadic interactions in the network with the aim of only including interactions that occur more frequently than expected if interactions occurred by chance. This approach, though appealing, has its methodological problems, which we discuss in detail in chapter 5. Thus it is safer to view association indices as just another means of weighting edges. As with filtering by a simple measure of co-occurrence, a rule of thumb for how hard to filter a network on this basis would be very useful, and we discuss this further in chapter 5.

Box 3.3

Examples of Association Indices

The aim of an association index is to improve on the simple count of group co-occurrence as a measure of the strength of association (edge weight in our language) between two individuals. If two animals are seen together three times, but then seem apart many times, should they be treated as being more closely associated than two others seen to gether twice, but never seen apart? No association index can be expected to make a perfect adjustment to imperfect data. An index that corrects for one bias may well introduce others. Thus there are many association indices that have been proposed over the years, each of which is likely to be of some use if applied to an appropriate data set.

Several useful association indices can be built on the following four measures: The association strength X, which is just the number of times a pair of animals (a and b, say) was observed in the same group; the number of times a was observed in a group, but not b, denoted Y_a; Y_b (similar, except b was seen); and Y_{ab}, the number of times both were observed but in different groups. The different indices using these measures simply combine them in different ways. For example, the simple ratio index (*SRI*) measures the fraction of times that a and b were seen together out of all the times either was seen:

$$SRI = \frac{X}{X + Y_{ab} + Y_a + Y_b}.$$

In situations where the observer is able to locate all individuals at each sampling point and record their associations (this will typically only occur in studies involving captive populations), the *SRI* gives exactly the same answer as the simple count of co-occurrence (X), albeit scaled between 0 and 1 (with a value of 1 indicating that the pair was always observed together and a value of 0 if the pair never associated).

Sampling biases may occur if the probability of locating an individual is dependent on the number and type of other individuals in a group, or many other factors that are related to patterns of association with other members of a group, such as reproductive condition (Cairns and Schwager 1987). For example, outside the breeding season we may be much more likely to see adult male and female deer in separate groups, but females and their calves will be more likely to be seen in the same group. In instances where there is a sampling bias to locate individuals when in separate groups, the half weight index (HWI), given by

$$HWI = \frac{X}{X + Y_{ab} + \frac{1}{2}(Y_a + Y_b)},$$

may be the most appropriate. In contrast, if the sampling bias is such that individuals are more likely to be located when together in one group, the twice weight index (TWI), given by

$$TWI = \frac{X}{X + 2Y_{ab} + Y_a + Y_b},$$

is the one to choose.

Alternatively, we may define social interactions based on patterns of association between local neighbors in a group (see chapter 2). In this instance we may be more attracted to the sociality index developed by Sibbald et al. (2005), which calculates the relative proportion of observations for which a particular pair of individuals were the nearest neighbors in any group.

In addition to correcting for biases within a sampling event, it may be possible to correct for other biases in the data. As we discussed in chapter 2, it is essential that we consider the independence of multiple sampling events. In instances where the sampling occurs more frequently than the rate of intergroup exchange, the samples will not represent independent observations and may greatly overestimate indexes of association. For example, in a study of the African buffalo Cross et al. (2005) recorded only 375 fission events over a 22-month period, but sampling occurred two to three times per week. To compensate for this Cross, Lloyd-Smith, and Getz (2005) proposed an association index they termed the fission decision index (FDI), which is the proportion of fission events involving both individuals in which they choose the same post-fission subgroup:

$$FDI = \frac{T_{ij}}{T_{ij} + A_{ij}}.$$

Here T_{ij} is the number of times individuals i and j were together after fission events, and A_{ij} is the number of times i and j separated during fission events. This index limits sampling to the point during fission and any subsequent fusion events, and thus controls for the sampling bias.

Finally, in a non-biological context, Newman (2001b) suggested that networks derived from co-membership of groups could have their edge weights reduced by a simple function of the group size. Thus if two individuals are seen together in a group of size g, one could add $1/(g-1)$ to their edge weight, not just 1. In an animal social network context, this adjustment would to some extent address the concern that two animals in a large group are less likely to have had a meaningful association than those in a small group.

4

Node-Based Measures

Now that we have a social network for our system, it is time to begin a more quantitative exploration of its structure, and to think in more detail about what this structure may tell us about the biology of our population. In this chapter and the next we will be tapping into some of the analytical tools and tricks developed by social scientists and others to pluck some order from the often-tangled mess that, at first glance, a social network can present. In this chapter we will consider several simple (to understand, if not to compute via mental arithmetic) but rather useful measures of the whole network structure, which are based on averages over the connection properties of the individuals. Inevitably there will be some variance in most of the measures at the individual level; it would be a surprise if there were not. It is then a natural line of enquiry to ask whether the distribution of these individual-based values over the network reflects some property of our population that might be interpreted in biological terms. Chapter 5 presents the relevant statistical methods to help us explore some of these issues. In chapter 6 we move on to slightly more sophisticated methods for teasing out whether a social network contains higher-level structures, involving several or many of the individuals, and in chapter 7 we explore methods for comparing two or more networks.

The main content of this chapter is an explanation of the meaning of some of the more commonly used descriptive statistics derived from network structures, and their use. We describe each metric, provide the mathematical equation required for its calculation, and then illustrate its potential interest in biological applications. While some of the measures may look daunting to the more mathematically challenged of us, there is no need to panic, as most can be easily calculated at the click of a button in standard network analysis software. It will soon become clear that progress in quantitative network analysis is practically impossible without the aid of a computer program or package and, as we have already mentioned in chapter 1, there are plenty to choose from. We have chosen here to concentrate on the UCINET package (Borgatti, Everett, and Freeman 2002, and see box 1.1), which in our experience offers a usable environment to get started with quantitative network analysis, and which gives access to several visualization programs. Most of the network properties introduced in this chapter can be calculated in UCINET.

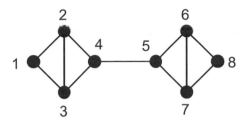

Figure 4.1. A simple 8-node network used to illustrate some quantitative measures of network structure. The edges that exist are all assumed to be undirected and of unit weight.

As with many interdisciplinary subjects that have been developed several times independently, network theory is rife with synonyms, which can confuse the unwary. We will attempt to stick to one set of terms (network rather than graph; node not vertex; edge not line or link) but to introduce some synonyms when an awareness of them is useful.

To aid explanation we will use a very simple toy network (figure 4.1) consisting of eight nodes (each representing an animal) and eleven distinct edges (representing a social tie between the animals at either end) to illustrate each structural measure and to introduce the means by which it is calculated and described. The edges in this toy network are both binary (an edge between two nodes either exists, in which case it has unit strength, or it does not) and undirected (an edge between two animals A and B implies A has a connection to B and vice versa). The extension of the measures of network structure we are about to explore to networks with more general edge properties is sometimes straightforward, but not always so. We will return to these issues in sections 4.7 and 4.8.

4.1 NETWORK EDGE DENSITY

We begin with a simple measure of the number of connections in a network, which may be useful to monitor when constructing, filtering, and comparing networks. All of the nodes in our toy network in figure 4.1 are in the same component (see chapter 3), since each is connected to all other nodes via a finite number of edges. However, the network is not "fully connected," as it is certainly not the case that all possible edges between the nodes exist. For a network containing n nodes, the maximum possible number of edges (with one edge between every distinct pair of nodes) is $E_{max} = \frac{1}{2} n(n-1)$. A quiet doodle with a few nodes should satisfy you that this is so. If the actual number of edges in a network is E, a first useful measure is the density of the network, ρ, which is just the fraction of possible edges that are present, and is given by

$$\rho = \frac{E}{E_{max}} = \frac{2E}{n(n-1)}. \tag{4.1}$$

In our toy network, $n = 8$ and $E = 11$, so $\rho \approx 0.39$.

So what does this tell us? Perhaps not a great deal on its own, but it is worth pointing out that in most social networks, ρ is much smaller than one (so that the network may be referred to as "sparse"). Thus most of the direct connections that could exist between pairs of animals don't in fact exist. In other words, the majority, and often the vast majority, of social connections holding together the population network are indirect, via intermediate agents. The network density will also need to be considered when we try to compare networks in chapter 7.

We now move on to a series of measures that are derived from the structure of the network edges, but which may be associated with individual nodes or, more rarely, edges. It is often the case that only the mean values of these measures are used to quantify network structure, but this ignores a good deal of available information which, if coupled to categorical data, might be more illuminating than a simple mean, as we will explore in the next chapter. We have therefore chosen to emphasize the single-node origin of these measures.

4.2 PATH LENGTH

The influence of indirect links in a network may be explored by calculating the distance, or path length, between a pair of nodes, counted in edges. Look again at the network in figure 4.1, and focus for the sake of argument on node 2. How far, in edges, is node 2 from node 1? There are many answers to this question, depending on which route we take, so to simplify matters we will concentrate solely on the *shortest* path, which will tell us how closely connected the nodes are at best. Following the tradition in network analysis, we will refer to the shortest path length as simply the path length (or occasionally the "geodesic" if the mood takes us) between two nodes. Then the path length between nodes 1 and 2 (and any other network neighbors directly connected by an edge) is 1. A useful shorthand for this is to write the path length as d, with a subscript listing the 2 nodes of interest; thus $d_{12} = 1$. Similarly, $d_{27} = 3$, since the shortest route between nodes 2 and 7 passes along the edges between 2 and 4, then 4 and 5, and finally 5 and 7. As a final example, $d_{14} = 2$; though there are two alternative shortest paths of equal length (via either node 2 or node 3), it is only the length itself that matters.

TABLE 4.1.
Numerical values of some of the simple node-based measures
of network structure evaluated for the toy network in figure 4.1.
The five columns represent the node label (i), the average path
length from that node to each of the others (L_i), and the clustering
coefficient (C_i), degree (k_i), and betweenness (B_i) of that node.

i	L_i	C_i	k_i	B_i
1	2.857	1.000	2	0
2	2.143	0.667	3	2.5
3	2.143	0.667	3	2.5
4	1.714	0.333	3	12
5	1.714	0.333	3	12
6	2.143	0.667	3	2.5
7	2.143	0.667	3	2.5
8	2.857	1.000	2	0
Mean	2.214	0.667	2.75	4.25

A simple individual-based measure of distance from other nodes is just the average distance from a node (node i, say) to all of the other $(n-1)$ nodes in the network. Denoting this as L_i we have

$$L_i = \frac{1}{(n-1)} \sum_{j=1}^{n} d_{ij}. \tag{4.2}$$

Let us pause here and make some notes. Pedants might note that the sum contains n terms, not $n-1$, but that's fine so long as we always choose to set the distance from any node to itself as zero. They might equally point out that L_i is a useless measure in a network with more than one component, since the distance from node i to nodes in other components is infinite, so L_i is infinite, and doesn't tell us very much. This is true, but there are ways around this problem, two of which we will mention shortly. Finally, lovers of unlikely names for things might like to note that the value of the summation in this equation is sometimes called the "farness" of node i. There are lots of "-nesses" in network parlance, one or two of which we will use a fair bit. You might like to amuse yourself by keeping an eye out for them, and perhaps noting which you think is the most ridiculous sounding "–ness" in the subject.

The individual values of distance to other nodes are not as frequently used in the networks literature as the average path length, to which we turn in a moment. Their reciprocal is the basis of a measure known as "closeness" (a fairly low-key "-ness" in our view). The merit of such measures can begin to be seen in the list of the L_i for the toy network in figure 4.1. These are shown in table 4.1.

The first thing to note in this list is the symmetry—nodes 1 and 8, nodes 4 and 5, and nodes 2, 3, 6, and 7 share the same value of L_i, as we should expect given that these nodes are completely equivalent to one another in the network, sitting in exactly the same local and global environment. This is, of course, merely a consequence of our choosing such a simple and symmetric network to play with, and is unlikely to be seen in a real-world animal social network. What is more interesting is that the relatively high values of L_1 and L_8 reveal what we can easily see in such a small network, that nodes 1 and 8 are rather peripheral nodes. Nodes 4 and 5 are clearly more central, and this is reflected (though not sharply) by the relatively low values of L_4 and L_5. This is first indication that simple measures might be useful in extracting differential roles and positions of individuals within a network, which is potentially very useful if these differences can in turn be related to biologically relevant factors.

The mean path length of a network, L, is simply the average of all the pairwise distances d_{ij} for the whole network, or equivalently, the average value of each of the node values L_i. Recall that we have already decided to set the distance from a node to itself as zero, so for an undirected network containing n nodes in a single component, there are $\frac{1}{2} n(n-1)$ distinct pairs to consider, so the mean path length L is

$$L = \frac{1}{\frac{1}{2}n(n-1)} \sum_{i<j} d_{ij} = \frac{1}{n} \sum_{i=1}^{n} L_i. \tag{4.3}$$

There are twenty-eight distinct path lengths that must be determined for our toy network, yielding a mean path length L of 2.214. This is a tedious calculation to perform even for an eight-node network, but it is a relatively easy one to compute using, for example, the UCINET package (where you should go to *network > cohesion > distance*).

L is referred to as a "global" network measure, since to find the distance d_{ij} between two arbitrarily chosen nodes i and j requires us to consider routes through the whole network. It is a useful measure since it gives us a feel for how close, on average, two members of the population are to each other (in a social sense). If the network connections are a good representation of the equilibrium social contact structure, for example, L would give an indication of how quickly a piece of information, starting from one arbitrarily chosen node, might be expected to spread through the entire population. Of course it is a much more difficult proposition to model in detail the dynamics of the spread of information through networks of a given structure. Such analyses have been performed for the case of epidemic diseases (e.g., Pastor-Satorras and Vespignani 2001) and rumours (Moreno, Nekovee, and Pacheco 2004), and have provided some of the recent success stories in network theory, yielding several counterintuitive yet important results. Yet with far less effort, the simple measure L gives us at least a feel for how easily different parts of a society may be connected. As a famous example of the use of L, the human population of the United States seems to have a path length of approximately six individuals, so that any two individuals can be connected via five intermediaries; this is the origin of the famous expression "six degrees of separation" (Milgram 1967). Another distance measure often quoted in the literature is the network "diameter" D, which is simply the largest of all of the d_{ij} values.

Before leaving path-length calculations we should, as promised, consider two of the methods for dealing with networks having more than one component. One approach (the one employed within UCINET) is to calculate path lengths only among "reachable" pairs of nodes or, in other words, pairs of nodes in the same component. For example, if we imagine the network in figure 4.1 with the edge between nodes 4 and 5 removed, we have a network with two components. We then proceed to compute the mean path length from each node to each of the others in its component. In this case we would find that the mean path length among reachable pairs would be 1.167. This is less that the value for the "full" network in figure 4.1, as we might expect. We have avoided infinities in our calculation, which is a good thing, but it is no longer quite so obvious that the mean path length is a global network measure in the way that it was. An alternative approach would be to calculate the harmonic mean path length L_{harm} via the equation

$$\frac{1}{L_{\text{harm}}} = \frac{1}{\frac{1}{2}n(n-1)} \sum_{i<j} \frac{1}{d_{ij}}.$$

There is no great mystery to this option, and for most situations L_{harm} and the arithmetic mean path length L would have very similar interpretations. Latora and Marchiori (2001) have suggested that $1/L_{\text{harm}}$ could be used as a measure of how efficiently a network could allow the movement of information. More pragmatically, the harmonic mean can simply be used as an alternative measure of path length, which happens to avoid infinities due to unreachable pairs. Harmonic means are always less than (or equal to) arithmetic means.

4.3 CLUSTERING COEFFICIENT

The clustering coefficient C is another measure of a network's average structure, and one that is complementary to the mean path length L. This is because it is derived entirely from *local* considerations of the network structure around each of the nodes.

Let us first calculate a clustering coefficient for our small toy network in figure 4.1 and then decide what it means. The first thing we do is pick a node (we've chosen number 7 this time) and identify the nodes in its "neighborhood," which are nodes 5, 6, and 8; the neighborhood contains all the nodes a distance of one edge away. Next we calculate the maximum number of edges that could exist among the members of the neighborhood, were they all directly connected to each other. This depends only on the number of neighbors; if there are k_i nodes in the neighborhood of node i, they could have at most ½ $k_i(k_i - 1)$ edges between them; so for node 7, this is ½ $3(3 - 1) = 3$. Then the clustering coefficient of node 7, written C_7 is just the fraction of these neighborhood edges that really exist in the network, which is ⅔, since two of the possible edges (between 5 and 6, and between 6 and 8) are there in figure 4.1, but the other (between 5 and 8) is not. Those of a geometrical disposition will have noticed that the number of edges in the neighborhood of node i is the same as the number of triangles, t_i, of which node i is a part. This leads to the rather neat expression for the node clustering coefficient as

$$C_i = \frac{2t_i}{k_i(k_i - 1)}. \tag{4.4}$$

The same process may be repeated for all n nodes in the network. The values for our toy network are listed in the third column of table 4.1. It should be mentioned in passing that the variation in the C_i follows that in the path lengths L_i, which might prompt us to infer that the node clustering coefficient is another measure of how central or peripheral the node is to the network. This is not the case in a general network; this should serve as a reminder of the danger of reading too much into simple illustrative examples. We will return to the relationship between path lengths and clustering coefficients a little later in the chapter.

The network clustering coefficient is simply the average of each of the node clustering coefficients:

$$C = \frac{1}{n}\sum_{i=1}^{n} C_i. \tag{4.5}$$

Every value of C_i, and therefore C itself, will always lie in the range $0 \leq C \leq 1$. If a social network has a large value of C (as many of them do), the implication is that many of one's social contacts are themselves directly connected. Thus the clustering coefficient is a measure of average local "cliquishness."

If using UCINET, you should be aware that there are several calculations that are associated with the keywords "cluster" and "clustering." To calculate clustering coefficients, go to *network > cohesion > clustering coefficient*. The "Overall graph clustering coefficient" quoted there is C. Also note that pendants (nodes connected to the rest of the network via only one edge) are ignored by UCINET in the calculation of C, rather than giving them a node clustering coefficient of zero.

An alternate way to arrive at the same formula for C_i is worth outlining here, as it emphasizes that clustering coefficients measure the extent to which one's

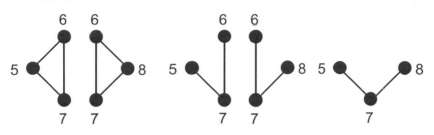

Figure 4.2. The triangles (left) and connected triples (right) used to calculate the clustering coefficient of node 7 in the toy network of figure 4.1.

immediate network neighbors are themselves neighbors. C_i may be calculated as the number of triangles connected to node i divided by the number of connected triples (defined as a single individual with edges connecting it to a pair of others, forming a "V") centered on node i. Node 7 in figure 4.1 belongs to 2 triangles (containing nodes 5, 6, 7, and nodes 6, 7, 8) and is the center of 3 connected triples (containing nodes 5-7-6, 6-7-8, and 5-7-8) so, as before, $C_7 = \frac{2}{3}$. Figure 4.2 illustrates these triangles and connected triples for node 7.

This method of defining clustering shows the importance of triangles in interconnecting agents in a network. The triangle is an example of a "motif," a small subsection of a network, such as a closed loop or small tree. Motifs have been used as the basis of an alternative method of quantifying network structure, particularly for directed networks. We will return to them in section 4.8 and chapter 7.

The clustering coefficient C is a network measure derived from local structure. In social animals this local structure may be due to active associations, such as phenotypic assortment, or associations between familiar individuals. Knowledge of C may contribute, for example, to our understanding of population susceptibility to the flow of information. For example, using a model of epidemic spreading on networks with a fixed number of edges but tuneable clustering coefficients, Newman (2003c) argued that even a weakly transmissible disease could saturate a susceptible population more quickly the higher the clustering coefficient.

It is worth pointing out at this stage that many of the data-collection protocols used to construct animal social networks so far have invoked the "gambit of the group" (chapter 2). An edge is drawn between each and every pair of individuals observed in each group, and the population network is built by accumulating these group-level clusters. In the jargon of network theory, networks constructed this way are "bipartite," in that a natural way to think of them is to define two types of node, representing individuals and groups, and to represent group membership with an edge running between an individual's node and a group node. Though this is an elegant approach, we have opted in this book to consider the so-called "one-mode projection" of any bipartite networks, with one type of node (representing individuals). These networks can then be visualized and analyzed in exactly the same way as any other network, including those derived from truly pair-wise interactions. The point we wish to make here is that completely connected groups have a clustering coefficient of one (they are perfect "cliques"), so the unfiltered network constructed from the accumulated data from many groups is likely to have an artificially high clustering coefficient. This effect will be countered by edge-filtering the network, as described in chapter 3 (and again in chapter 5), since then most of the intra-group edges from particular group will be lost. The bias introduced into a calculation of C by invoking the gambit of the group can, and probably should, be monitored by the use of a randomization test; this will be discussed more fully in chapter 5.

4.4 DEGREE

The degree of node i in a network, written k_i, is simply the number of edges connected to it. We have already used this notation when evaluating the clustering coefficient of a node. The node degrees for the simple network in figure 4.1 are listed in table 4.1. There are many features of node degree, both individual and collective, that are worthy of consideration by the budding network analyst. The simplest among these is the mean degree k, which is the average of the individual node degrees:

$$k = \frac{1}{n}\sum_i k_i. \tag{4.6}$$

In our example, $k = 2.75$. The degree of an individual tells us the number of different social connections an individual has and can be easily calculated in UCINET at *network > centrality > degree*. The mean degree is a very simple measure of how well connected a node is, on average. Note that since each edge contributes to the degree of each of the two nodes it connects, the mean degree of any network with n nodes and E edges can be written as $k = 2E/n$.

The degree distribution of a network, $P(K)$, is defined as the fraction of nodes that have degree K (or, just as often in the literature, greater than or equal to K). The nature of the degree distribution of many large networks has been the subject of a huge amount of interest in network theory (see, e.g., Boccaletti et al. [2006] for a review), and can be an important indicator both of network topology and the dynamics of the propagation of information, including rumors and epidemics, on the network. Imagine, for example, that one individual produces an innovative foraging technique such as washing sweet potatoes that was observed in Japanese macaques (*Macaca fuscata*) (Itani and

Nishimura 1973). It seems reasonable to assume that the degree of the innovator could be an important determinant of whether the innovation will spread in the population or not. Individuals with high degree may easily transmit information to others.

Finally, there has been a fair amount of interest in the nature of degree correlations (the tendency for well-connected individuals to be directly connected to other well-connected individuals) in human social networks. We will return to this topic in chapter 6.

4.5 OTHER CENTRALITY MEASURES: NODE AND EDGE BETWEENNESS

The degree of a node is a very simple measure of its "centrality"; nodes with many neighbors are "well-connected" and tend to occupy central positions in a network, and those with very few will be on the periphery. Many of the network measures developed by sociologists (see, e.g., Wasserman and Faust 1994; Scott 2000) that are appealing to the biologist attempt to discriminate between individual nodes, often with the aim of determining which individuals (or categories of individual) are the most important or "central" to the network in terms of holding it together as a single component, being a potentially important broker for the transmission of information, and so on. There are thus many measures of centrality, often more sensitive than node degree, that are well worth exploring; the UCINET package offers plenty of options at *network* > *centrality*. As an aside, one of the reasons that spring embedding (chapter 3) is such a useful visualization tool is that it tends to place high-centrality nodes in the center of the graph and low-centrality nodes at the periphery.

We will concentrate on two closely related centrality measures. The "node betweenness" of an individual node i is defined as the total number of shortest paths between pairs of nodes (other than i) that pass through i, and is written B_i. This is a tricky quantity to calculate, but it is relatively easy to see that the number of shortest paths passing through a node is indeed a decent measure of the centrality of the node. At first sight, it would appear that the node betweenness should be strongly correlated with node degree, and often it is, but this need not be so. In our toy network in figure 4.1, all the nodes have degree 2 or 3, but nodes 4 and 5 have a much higher node betweenness than the others (see the final column in table 4.1) because all of the paths connecting nodes on the left to those on the right must pass through these nodes.

Our simple model network reveals a general truth that the node betweenness is often a more sensitive discriminator of the relative positions of nodes in a network than, for instance, the mean distance to other nodes (L_i). By computing the node betweenness we have a measure of which individuals in a population are the key "players" in terms of information flow in a population. For example, in an investigation on bottlenose dolphins (*T. truncatus*) living in Doubtful Sound, New Zealand, Lusseau and Newman (2004) identified animals with relatively high betweenness in the population that were located at the boundaries between sets of socially clustered individuals (or "communities" as they are known in the networks literature—see chapter 6). The authors suggested that these individuals may play a particularly important role in maintaining the social cohesion of the population. One can imagine a situation where similar information could help us target conservation efforts and management strategies.

A closely related structural property, but one based on edges rather than nodes, is the edge betweenness. This measures the number of shortest paths that traverse a given edge. In the toy network in figure 4.1, the edge connecting nodes 4 and 5 has by far largest edge betweenness (16, with the next highest being 7.5—can you work out which edges these are?). Edge betweenness has been used as the basis of determining intermediate level or "community" structure in networks, as we will see in chapter 6.

There are many other measures devised to unravel details of an individual's role in a social network. One we mention in passing is the "node reach centrality," which is the number of other nodes a distance of q or less away. So for $q = 1$, the reach of a node is equal to its degree. Flack et al. (2006) uses a node reach with $q = 2$ to quantify network structure in pigtailed macaques (*M. nemestrina*). She argues that this measure, which counts the neighbors of one's neighbors, could be a useful way to capture the propagation of behavioral traits across a network. It may be that some of the measures we have omitted will prove in the future to be of great use in the characterization of animal social networks. However, we are confident that by the time you are considering moving to other measures, you will be happy enough with the notions and terminology of network analysis to find and interpret them for yourself. The book by Wasserman and Faust (1994) gives an excellent and clear account of many such measures for those who are ready.

The remainder of this chapter is dedicated to two topics. The first is an introduction to some of the model networks that are well understood and help us decide what type of network we have. The second is a brief exploration of how we must rethink some of the concepts and measures outlined above if our network edges are either weighted or directed.

4.6 SOME MODEL NETWORKS

An objection that could easily be raised to the measures we have presented so far is that, though they are quantitative, they don't really give us much of a feel for what type of network we have. Having gone through the protracted business of designing an investigation, observing the animals, constructing a network and getting to grips with some software, it would good to know whether your values of L, C, and k were entirely mundane or earth-shatteringly excit-ing. Should you brag to your friends and colleagues that your iguanas have a clustering coefficient of 0.35, or is this something to keep to yourself?

It turns out that this is a rather important question, and one that must be addressed before we can start bandying about too many figures. In general, the most satisfactory answer is to compare your measures with those calcu-lated for other networks. Chapter 5 introduces some techniques for compar-ing network measures to those constructed from randomized versions of the same data, which serve as null models. Chapter 7 contains a discussion on how to compare one animal social network with another. As we will see in those chapters, the choice of comparison network can require some thought, and the detailed comparison will require some computation. Before we get to those niceties, though, there is merit in comparing the structural proper-ties of a single empirical animal social network with two rather simple and extremely useful model networks. These are random networks (often called "random graphs" by mathematicians) and regular networks (also known as "regular graphs" or "lattices"). Each can be constructed with the same num-ber of nodes n and edges E as the original network (and therefore the same mean degree k). These model networks are very well studied and understood, and act as simple markers with which your network can be compared. A very brief account of the relevant properties of random and regular networks is in order.

Random Networks

There are many types of random networks, but the type we are interested in here contains n nodes and E edges, with each of the edges placed between two nodes chosen at random. Such networks are often referred to as "Erdös-Rényi random graphs" (Erdös and Rényi 1959). To be consistent with our animal social networks we can stipulate that no self-edges are allowed, nor are any duplicate edges; all edges must connect a distinct pair of nodes.

So what are typical values of the average path length L and clustering coef-ficient C for a random network with n nodes and E edges? C can be estimated by noting that the clustering coefficient of a given node measures the fraction of its connected triples that are, in fact, a part of a triangle (see figure 4.2). For our random network this is equal to the probability that the edge turning a triple into a triangle is present, which is the same as the probability that any chosen edge is present, and this is simply the edge density $\rho = E/E_{max}$. The same result holds for each of the node clustering coefficients, so the clustering coefficient of our random network is

$$C_{rand} = \frac{E}{E_{max}} = \frac{2E}{n(n-1)} = \frac{k}{n-1}, \tag{4.7}$$

where the last step follows because $k = 2E/n$ for any network. This result shows that, given that most of the networks we are interested in have low edge density ρ, the random network with the same number of nodes and edges has a rather small clustering coefficient. It also indicates that, for example, calculating very different clustering coefficients for two networks might not be a source of great biological interest if the network with lower clustering has many more nodes; the number of nodes may well be the primary source of the difference.

The mean path length of a random network takes a little more calculation, which we will spare you and simply quote (see e.g., Bollobás 1985):

$$L_{\text{rand}} = \frac{\ln(n)}{\ln(k)}, \tag{4.8}$$

where ln denotes the natural logarithm. This might not look very helpful yet, but the key point is that though L_{rand} increases with network size, it does so rather slowly. Thus random networks with more than a few nodes have a relatively small mean path length; this is an intuitively reasonable result, as can be seen by looking at the example random network in figure 4.3c. If edges are placed between randomly chosen nodes, it is very unlikely that there will be any nodes a long way from others. Instead there are likely to be shortcuts which keep the mean path length small.

Mathematicians have discovered many features of very large random networks (see Bollobás 1985), one of which we mention here as an aside. If we build a random network by progressively adding more and more edges, and look at the number of components in the network as we go, a curious thing happens. Until $n/2$ edges have been added (i.e., before the mean degree k has reached 1), the network always consists of many small components. Once there are more than $n/2$ edges added, suddenly a "giant connected component" (GCC) appears, containing virtually all of the nodes with a few stragglers outside. In other words, the network goes through a "percolation threshold." The interested reader should take a look at one of the excellent reviews of complex networks in the physics literature, such as those by Albert and Barabási (2002), Newman (2003a), and Boccaletti et al. (2006) if they wish to understand more of the intriguing statistical phenomena exhibited by random (and other) networks.

Why mention such arcane points in an introductory text on animal social networks? Well, even though these results, and many like them, are derived for large random networks, and we are primarily interested in analyzing relatively small, nonrandom networks, they still have relevance. If we construct a social network using point sampling, for example, and discover that there are several disconnected components, it is tempting to think that we have learned something important about segregation in our population. If, however, we find that

a   b   c

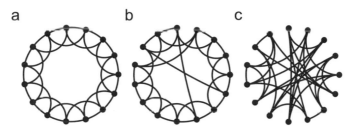

Figure 4.3. An example of a regular network (a), a random network (c) and a small world network (b). In each case the degree of each node is 4. (b) and (c) are formed by randomly rewiring more and more of the edges in (a). After Watts and Strogatz (1998).

the mean degree for our network is less than, or even close to, 1, we should be aware that the existence of several components is almost inevitable, so we should endeavor to take more samples before getting carried away with the meaning of our results (see chapter 3).

Regular Networks

In a regular network, each node occupies an identical neighborhood. There are several ways to achieve this; one such is shown in figure 4.3a. All nodes in the figure have degree 4, with their edges connecting them to the two nearest spatial neighbors on either side. Again, the calculations to find L and C for this particular network are possible, but not particularly illuminating. They yield the results (Watts 1999) $C_{\mathrm{reg}} = \frac{1}{2}$ and

$$L_{\mathrm{reg}} = \frac{n-1+k}{2k}. \tag{4.9}$$

More generally, it is useful to observe that regular networks have a high (often higher than ½) clustering coefficient that does not change with network size, and a path length that is given (at least approximately for any network with n appreciably larger than k) as

$$L_{\mathrm{reg}} \approx \frac{n}{2k}. \tag{4.10}$$

This value rises much more quickly than L_{rand} with increasing n, so large regular networks tend to have relatively high values of clustering coefficient, and high values of mean path length. Again, the latter is clearly illustrated in figure 4.3a; the path length between nodes on opposite sides of the ring is large because it involves taking many small steps around the perimeter.

Small-Worlds Networks

Since the late 1990s there have been many contributions to the field of network characterization from statistical physicists, adept at turning their hands to the development of efficient computer algorithms for complex systems and to finding analytic results in "the thermodynamic limit" (in other words, for networks with infinitely many nodes). An important paper by Watts and Strogatz (1998) introduced a toy model (figure 4.3) that elegantly transformed a regular network into a random network via the tuning of a single "re-wiring" parameter. Their model showed that just a few long-range random connections could produce "small-worlds" networks with short mean path length L, not much larger than that of a random network with the same number of nodes and edges, yet retaining the large clustering coefficient C of the equivalent regular network. This was an important result, since as Watts and Strogatz (1998) showed, the real networks of many apparently disparate systems exhibit the "small-worlds" phenomenon.

Animal social networks also exhibit properties that are consistent with the small-worlds phenomenon (at least in regard to the values of their mean path length and clustering coefficient). Systems for which this has been reported include bottlenose dolphins (Lusseau 2003; Lusseau et al. 2006) and Trinidadian guppies (Croft, Krause, and James 2004a). This is of potential interest in biology because small-worlds may facilitate the rapid transmission of social information and socially transmitted diseases through highly clustered networks (Watts and Strogatz 1998; Latora and Marchiori 2001) such as those found in animal populations. However, we should not get too carried away here. As pointed out by Newman (2003c) and others, it is actually quite difficult to find real networks for which the mean path length is closer to the value expected of a regular network than that of a random network; in most cases there are at least a few edges that form "shortcuts" and yield a relatively low value of L. Thus simply finding that L is slightly greater than would be expected for a random network with the same number of nodes and edges is not, in itself, a very interesting result. (Perhaps confusingly, this feature alone is sometimes referred to as the "small-worlds effect.") Finding that the clustering coefficient is also much greater than for the corresponding random network is more interesting, though caveats we have already mentioned will apply if the network was constructed via the gambit of the group.

Scale-Free Networks

A similar note of caution should surround the identification of one's animal social network as exhibiting "scale-free" properties. The term arises from the shape of the degree distribution $P(K)$. For an Erdös-Rényi random network, and the toy

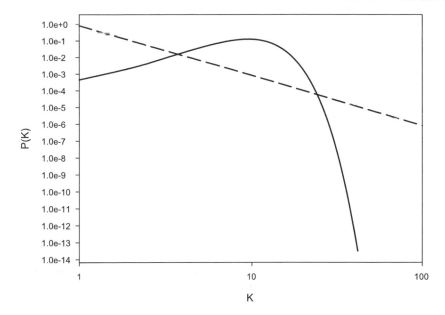

Figure 4.4. Illustrative degree distributions, plotted on log-log scales. Solid line is the Poisson distribution for a mean of 10. Dashed line is a power-law distribution, with exponent $a = 3$. The power-law distribution is normalized to make the cumulative probability at $K = 1000$ equal to 1.

model of Watts and Strogatz (1998), $P(K)$ is binomial, or Poisson-distributed, in the large-n limit. For such distributions the strong maximum associated with the mean degree k sets a scale for the distribution in that we can easily tell whether a particular value of K is to the left or right of the peak, and therefore whether it is relatively low or high (see figure 4.4). By contrast, the networks derived from many very large real systems, including human social networks (such as e-mail messages—Ebel Mielsch, and Bornholdt [2002]) and technological networks (such as the Internet—Faloutsos, Faloutsos, and Faloutsos [1999]) are found to have so-called power-law degree distributions, for which $P(K) \sim K^{-a}$ (a is a positive constant) over a wide range of values of K. A plot of such a power-law distribution on log-log axes is a straight line, with slope $-a$ (figure 4.4). Along this line there is no value of K that stands out as being special; there is nothing to set the scale. For this reason, networks with power-law degree distributions are referred to as being scale-free.

There has been an explosion of interest in scale-free networks and their properties, not least because they have been shown to exhibit qualitatively different information- (and disease) carrying properties, and resilience to the removal of nodes or edges, compared to random networks (Pastor-Satorras and Vespignani 2001). The key feature that gives rise to many of the interesting properties of scale-free networks is the positive skew of the degree distribution, which is very pronounced; there are many more nodes with very large degree than would be expected for a Poisson-distributed network, and these "superhubs" can have a very pronounced effect on many network properties. Such findings make it very tempting to look for the signature of scale-free behavior in animal social networks by plotting $P(K)$ against K on log-log axes and looking for a straight line over some of the range of K-values. In truth, however, most animal populations used to date to construct social networks have been far too small to ascribe the same properties as seen in, for example, the Internet, for which the degree distribution is scale-free over at least three orders of magnitude of K-values (Faloutsos, Faloutsos, and Faloutsos 1999). (As an aside, it should be said that even very large networks cannot be perfectly scale-free in their degree distribution. For small K there is often some deviation from a power law. At the large-K end of the distribution, it is clear that for any network of finite size, there must be an upper limit to the degree of a node, which is the number of other nodes. So in fact, even very large "scale-free networks" are only scale-free over a finite range of K. Our point is that if that range spans several orders of magnitude in K, the assignation may have some merit; if the range is, say fifty, it doesn't.)

All of this does not mean that there is nothing to be learned from plotting and analysing a degree distribution—indeed we would encourage plotting the distributions of all manner of variables such as recapture frequency, group size, path lengths, and node betweennesses in order to try to understand the observed network structure. Figure 4.5 shows degree distributions for five populations (each of size 100 or so) of small freshwater fish. There may well be something to be learned from the similarities and differences in these plots, and it is interesting to note that the distributions have a similar shape, with a positive skew, to that reported for bottlenose dolphins (Lusseau 2003). However, the range in K is simply too small to assert that there is no natural scale to the degree distributions, or that the most connected animals are "superhubs" that are bound to have a key role in some process or other.

Our message here is a simple but important one. The field of network theory is exciting and rapidly developing, with contributions from people in a wide range of fields. There are frequent developments in analysis of structure and dynamics, some of which may turn out to be of enormous use in animal biology. However, we must show some restraint and read the "small print" of these developments before we jump in and use them. It is, for example, a foolhardy activity simply to adopt the language and methods used to describe very large networks, and to use the statistical results derived from them, without paying due regard to whether this is meaningful in the case of a few tens or hundreds of animals.

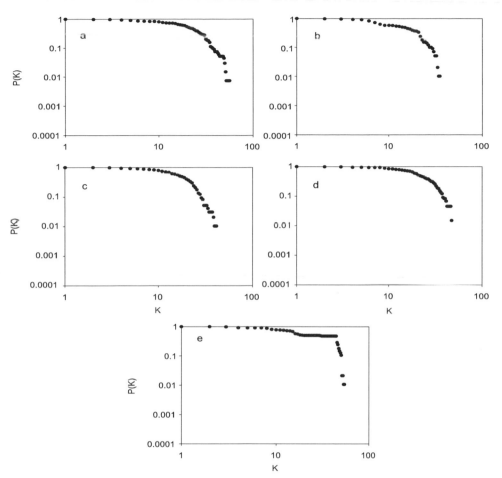

Figure 4.5. Cumulative degree distributions (the fraction of nodes with degree at least equal to K) for four populations of Trinidadian guppies (a–d) and a population of three-spine sticklebacks (*Gasterosteus aculeatus*) (e). From Croft et al. (2005).

4.7 NETWORKS WITH WEIGHTED EDGES

The reader just getting started with networks, or one who has for sure an undirected, unweighted (either in fact or by choice after filtering) animal social network can skip now to chapter 5. For the remainder of this chapter we will be looking at how the arguments and measures set out above must be rethought if we wish to take non-binary edge weights into account or if the social network you have constructed has directed relationships.

We have already seen in chapter 2 that there are good reasons to expect that an animal social network will have weighted edges, reflecting the fact that not all pair-wise interactions in a population will be of equal strength (or indeed, even of the same sign). Though some of the measures we have discussed in this chapter can be recast when the edges are weighted, it is currently not clear how to interpret these values. Thus, to date most analyses of animal social networks have used filtering (sometimes via the application of association indices) to construct a binary network, amenable to simple quantitative analysis. This trend may reverse if a better understanding of real and model weighted networks is forthcoming; the measures presented below are relatively recent in origin, and much less well tested than their binary network equivalents. Boccaletti et al. (2006) contains an excellent summary of weighted network measures. We note that a recent paper by Lusseau, Whitehead, and Gero (2008) strongly advocates the use of weighted measures for the analysis of animal social networks.

A weighted network may be represented by an $n \times n$ weight matrix $\mathbf{W}$, whose element W_{ij} represents the weight of the edge connecting nodes i and j. (For a binary network, all weights are either 0 or 1.).

Weighted Degree: Node Strength

The node strength

$$s_i = \sum_{j=1}^{n} W_{ij} \tag{4.11}$$

measures the total weight of the edges connected to a node, combining the degree of a node with the weight of all its edges. It is the weighted equivalent of the node degree, and parallel measures such as node strength distribution function $P(s)$ have been reported in some analyses of non-animal networks (Barrat et al. 2004).

Weighted Clustering Coefficient

Several methods of including edge weights in the calculation of clustering coefficients have been proposed, some of which are reviewed in Saramäki et al. (2007). A promising variant is a natural modification of equation 4.4 (Onnela et al. 2005). Instead of a simple count t_i of the number of triangles involving node i, we assign a weight to each triangle, based on its edge weights. This is most easily demonstrated with an example. Recall that the clustering coefficient of node 7 in figure 4.1 was found to be $C_7 = 2/3$, with the relevant triangles shown to the left in figure 4.2. Now let us suppose that the edge between node 7 and 8 has weight ½, and all the other 10 edges in the network have weight 1. (The edge weights must be scaled to the largest value in the network to keep all the node clustering coefficients between 0 and 1.) Now we define the weight of a triangle as the cube root of the product of the three edge weights; thus for the first triangle in figure 4.2, the weight is $(1 \times 1 \times 1)^{1/3} = 1$, but for the second it is $(\frac{1}{2} \times 1 \times 1)^{1/3} \approx 0.794$. The weighted clustering coefficient of node 7 is found by adding the triangle weights, instead of counting the triangles, so $C_7^W = \frac{2}{6} \times (1 + 0.794) = 0.598$, slightly lower than the unweighted value $C_7 = \frac{2}{3}$, as we might expect. Of course, node 7 is not the only one affected by the change in this single edge weight. Similar calculations reveal that $C_6^W = 0.598$ and $C_8^W = 0.794$. All other nodes (not connected to the affected edge) have the same values as before, so overall the network now has a mean weighted clustering coefficient of $C^W = 0.624$.

Weighted Path Length and Betweenness

The natural way to weight path lengths is to introduce an edge length δ_{ij} for each existing edge that depends on its weight. A simple choice is $\delta_{ij} = 1/W_{ij}$, which makes a path through a heavily-weighted edge shorter (and therefore more important, in this context). The shortest path between two nodes is then the path with the smallest total edge length. This complicates things, as now it need not be the case that the shortest path contains the fewest edges. For example, the shortest path between nodes 2 and 5 in figure 4.1 is 2, passing from node 2 to 4 to 5. However, if $W_{24} = \frac{1}{5}$ and the remaining edges have unit weight, $\delta_{24} = 5$, and the weighted path from node 2 to 4 to 5 has edge length 6, and is "longer" than the weighted 3-edge path via node 3, which has edge length 3.

Algorithms exist to calculate weighted shortest paths (see for example Newman [2001a]). From these it is possible to compute modified betweenness values, taking the edge lengths into account, though again their use and interpretation has not yet become a matter of routine.

4.8 NETWORKS WITH DIRECTED EDGES

By contrast with weighted edges, there is a long history in the social sciences of considering networks whose edges are directed. From the outset it has been understood that human interactions are not always reciprocated. Thus many of the methods developed to characterize human social networks have been designed with directed edges very much in mind (see the books by Scott [2000], Wasserman & Faust [1994], and Carrington, Scott, and Wasserman [2005] for example). Computer packages written for or by social scientists, such as UCI-NET, are designed to cope with them. In terms of the measures and concepts introduced in this chapter, the directedness of edges forces us to consider some extra variables.

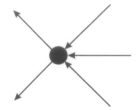

Figure 4.6. Node degree in directed networks. This node has an in-degree of 3 and an out-degree of 2.

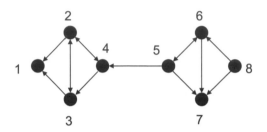

Figure 4.7. A simple 8-node directed network based on the undirected network of figure 4.1.

Degree and Components

One obvious difference introduced by the directedness of edges is that each node now needs two numbers to count the edges it is attached to: an in-degree k_i^{in} denoting the number of neighbors with an edge to node i, and an out-degree k_i^{out} that is the number of edges leaving node i (see figure 4.6).

So what difference does this make to our simple measures of network structure? Perhaps this will be easiest to see through another simple example. In figure 4.7 we have invented a second toy network, closely related to the undirected network we used before, but this time the edges are directed. Some edges point only one way (from node 2 to node 1 for example); others are reciprocated (node 2 is connected to node 4 and vice versa).

Let us evaluate the in- and out-degrees for each of the nodes. The results are listed in the 2nd and 3rd columns of table 4.2. We see that the in- and out-degrees are not generally the same for each node, but the mean values are equal. This is always true, as each edge must "start" and "finish" somewhere. The value of the mean degree is just the number of arrows divided by the number of nodes, or $14/8 = 1.75$ in this case.

The directedness of the edges means that even the identification of network components is less straightforward than before. Of course, a quick glance at

TABLE 4.2.

Numerical values of some of the node-based measures evaluated for the directed toy network in figure 4.6. The six columns represent the node label (i), then the out-degree, in-degree, path lengths from and to that node, and betweenness.

i	k_i^{out}	k_i^{in}	L_i^{out}	L_i^{in}	B_i^D
1	0	2	—	2.667	0
2	3	2	1	2.2	3
3	2	2	1.333	2.2	2
4	2	2	1.333	1.8	9
5	3	1	1.667	1.5	8
6	2	2	2.333	1	5
7	0	3	—	1	0
8	2	0	2.857	—	0
Mean	1.75	1.75	1.754	1.767	3.375

figure 4.7 would suggest there is only one component, but this is not quite the case. Consider node 1. Its out-degree is zero, so there is a sense in which it is isolated from the rest of the network; any information starting on node 1 cannot reach any other part of the network. Similarly, the in-degree of node 8 is zero, so no information moving through the network could ever reach node 8. To minimize confusion over this, it is useful to define, for each node, an "in-component" that is the set of nodes from which that node can be reached via a finite number of edges, and an "out-component" that is the set of nodes that can be reached from that node. So for example, the in-component of node 3 in figure 4.7 contains all the nodes except 1 and 7, while its out-component contains nodes 1, 2, and 4. Similar care must be taken when looking at other degree-based measures such as degree distributions and correlations.

Path Length and Betweenness

It should be clear from the above that not all pairs of nodes will be connected, even in a simple directed network like the one in figure 4.7, and that the distance from node i to node j need not be the same as the distance from j to i. Thus we must calculate these two path lengths separately. It also makes sense to avoid infinite values by considering only "reachable pairs" when calculating path lengths. In analogy with the node path lengths L_i of section 4.2, we may define L_i^{out} to be the mean distance from node i to members of its out-component, and L_i^{in} as the mean distance to node i from members of its in-component. The results are listed in table 4.2 for our toy directed network. A dash indicates that there are no paths to consider. Note that the mean values of these two quantities over all nodes are not equal in this case, and nor are they the same as the mean value of path length ($55/28 = 1.964$) for the network as a whole. These discrepancies only arise because we are ignoring different numbers of nodes or paths in each case; L_i^{out} is the average of 6 numbers, L_i^{in} of 5, and the mean path length is averaged over the twenty-eight individual path lengths between reachable pairs.

The node betweenness does not need to be decomposed into "in" and "out" values, since it is a count of shortest paths passing through a node. The paths must of course be the directed paths we have considered in computing path lengths. The values of node betweenness for our toy network are listed in table 4.2. Note that nodes 4 and 5 have the highest betweenness values, as they did for our closely related undirected toy network.

The interested reader should note that in the social sciences various measures of prominence have been developed specifically to analyze directed networks (Wasserman and Faust 1994). There are various measures of "prestige," which are based on a node's in-component. Similar measures on the out-component of a node are again termed "centrality" measures.

Clustering and Motifs

The situation with clustering coefficients is slightly more tricky. Though UCI-NET, for example, will compute a clustering coefficient for a directed network, it is not entirely obvious what it is measuring. As pointed out by Dorogovt-sev and Mendes (2003), "the notion of clustering [as presented here] is only well defined for undirected networks." Indeed, the notion of a "clustering co-efficient" has not really been a part of tradition in social network analysis, though closely related concepts such as transitivity certainly have been (see e.g., Wasserman and Faust 1994). As an alternative route to characterizing cliquishness and similar concepts in directed social networks, some authors have turned to analyses based not on simple metrics but on the identification and enumeration of various small structural fragments in a network (see, for example, Faust and Skvoretz [2002]; Milo et al. 2004; and chapter 6). Such small structural fragments of networks are known as "motifs." Examples include directed edges arranged among three nodes; the relative abundance of these carries essentially the same structural information for directed networks as does the clustering coefficient for the undirected case. The analysis of these motifs, and their frequency of occurrence in a network, is an interesting avenue of research for those developing methods to analyze network structure. It is easy to see that not all motifs will be present in all networks. For example, in a network of monogamous heterosexual contacts, there are no triangles of edges at all, though the more imaginative reader might be able to think of a series of personal arrangements that could give rise to various other network motifs. We will turn to motifs again in chapter 7, where we consider their use in network comparisons.

4.9 CONCLUDING REMARKS

The message to take from this chapter is that there are plenty of structural metrics available that can give a first quantitative measure of an animal social network. These measures are most straightforward to use when the network edges are both unweighted and undirected, though there are plenty of extensions available to use if your network does have directed or weighted edges, or both. Simple network models, such as random and regular networks, exist that can easily be used to make a preliminary assessment of whether average values of path length and clustering coefficient computed for your network are unexpectedly high or low.

5

Statistical Tests of Node-Based Measures

So we have collected our relational data, drawn our network, and used it to calculate some descriptive statistics. What next? As biologists we should be interested in what the social network can tell us about the sociobiology of our study species, and to this end we need to have some confidence that our observed network patterns stand up to statistical analysis. There is much that can be done in this regard, but also much still to do. The very aspect of networks that makes them so appealing, that they represent all relevant relations in one construct, is the aspect that makes them slightly trickier than other data sets to analyze. However, there is no need to despair, merely to apply some common sense in deciding how much of your quantitative network analysis is likely to be statistically robust.

Some of the features that can make a difference in how easy it is to test patterns in network data can be identified by the following questions:

- Do we have a relatively small (tens of animals not hundreds) system that we have been able to study continuously, having strong confidence we have observed all interactions?
- Is each "interaction" observed as such, or inferred from "association," via co-membership of a group, for example?
- Do we have replication?
- Can we manipulate the system to test any conclusions?
- Are we trying to compare node values of a small number of individuals, or representative measures such as means over categories (young males versus older females, say)?

Both this and the next chapter are concerned with statistical testing of patterns in a single social network. Often the questions we are trying to address, or the type of biological information we will be trying to extract, from a network will appear to be rather similar in these two chapters in that we will often be trying to relate patterns we find to attributes of nodes or groups of nodes. However, there is an important methodological difference in how we are going to go about the job. In chapter 6 we will look at the structure of the network as a whole, and ask what the topology of all the connections tells us about the relations that it represents. In this chapter we are going to see what might be learned about a system from the node-based measures of the type introduced in chapter 4. We may have observed that males have higher degrees than females, or that the clustering coefficients of females are greater than those for males, but how do we test whether these trends are statistically significant? We will explore what we can learn about the values of node-based network measures, their distributions, and how the measures are distributed among categories. Some of the issues highlighted along the way will be specific to the analysis of node-based measures; others will be slightly more general.

Before we go on, we should perhaps mention two rather important factors that potentially make life much easier for the network practitioner, namely *replication* and *manipulation*. In this chapter (and the next), we will be assuming that we have precisely one network representing social interactions, associations, or proximities, and will consider what we can do to extract useful data from it. It almost goes without saying that if we have replication at the level of the population (i.e., we have measured the same relations across multiple networks), then it may be possible to make statistical inferences by comparing network measures between contexts or to comment on the general trends in the networks. Chapter 7 addresses network comparisons in more detail. In addition, it helps greatly if biological conclusions from a network analysis can be backed up by empirical tests. For example, we may have inferred that two animals are seen together more frequently than we would expect by chance, and predict that this association represents a preferred social relationship based on active choices by the participants. This might be tested experimentally by giving one of the individuals a binary choice between associating with its observed partner or another with whom it has rarely associated. In a similar vein, if we are operating in a semi-natural environment and find that segregated habitat use seems to offer the most likely explanation for structural heterogeneity in our network, we might be able to manipulate the spatial structure and test our predictions on the network constructed under different ecological constraints. Perhaps due to the large effort usually required to gather relational data, rather few studies presented so far include either replicated networks or much in the way of verification via experimental manipulation.

5.1 AN EXAMPLE

Let us begin this chapter by choosing a scenario where just about everything is as straightforward as it could be. The most desirable starting point for a statistical analysis of any network is one where we are confident that, as far as is possible, we have constructed "the" network for our system, which correctly codifies the web of social interactions or associations of the type we are interested in. Then at least we can rule out any problem with sampling errors, or biases in the network due to variance in the number of times an animal is observed, and so on. In this (fortunate and rare) scenario, it seems reasonable to assume that we are at liberty to analyze any network measure we like, and to try to relate it to some biology of interest. We might look at the value of individual node measures, such as the degree k_i, the path length L_i, the clustering coefficient C_i, the node betweenness B_i (see chapter 4 for definitions of these terms), or perhaps at the means of these measures over given categories or the whole population. We might look for correlations between the distribution of node values and some ecological, behavioral, phenotypic, or genetic information we have collected about the individuals. (These are the sorts of things we mean by "node attributes.") Equally we might look to see whether the node values are consistent with those from an appropriate null model. For example, we may want to probe whether the observed network measures differ significantly from what we would expect if all interactions or associations between individuals occur randomly. Finally, we might do none of the above and, unencumbered by what others have done, strive to find a new measure or technique that most succinctly captures the biology represented in the network.

So let us explore how some of this might be achieved by returning to the entirely fictitious network we drew and discussed in chapter 1. Given that we invented the species (*Commenticius perfectus*) and everything about it, we will assume for now that our network is infallibly correct. Figure 5.1a shows the network again, with an animal's sex indicated by node shape and its body length (or whatever single continuous trait you prefer) indicated by node size. What can we say about the social structure of our population now that we know how to quantify its network? Well, we could look at mean measures of L (3.17) and C (0.35) and use the formulae of section 4.6 to compare these with values ($L = 2.39$ and $C = 0.175$) expected from a random network with the same number of nodes and edges. We might then infer that we have a "small-worlds" network (see section 4.6), and we would at least be on our way.

We might also make use of the fact that computer packages such as UCINET often provide summary statistics such as the standard deviation and skew of the various structural measures, which might again be used to give us confidence that we understand something about how the measure is distributed across our network.

But let us be a little more ambitious. Let us suppose that we study *C. perfectus* in two consecutive field seasons. In the first season we observe the animals and derive the social network in figure 5.1a. The following season we look at the same animals (they all survive, of course, and remain at our field site) and monitor the number of parasites on each animal. The easiest animals to catch are those distracted by how much scratching they are doing. On our first day we capture three individuals; we find that animal K has the most parasites, followed by F, then O. We then notice that in the previous season's network, K has a high node betweenness (the highest in fact), with that of F rather smaller and O smaller again. Then we get excited—perhaps we have found a network measure that predicts future parasite load? And moreover, perhaps the fact that

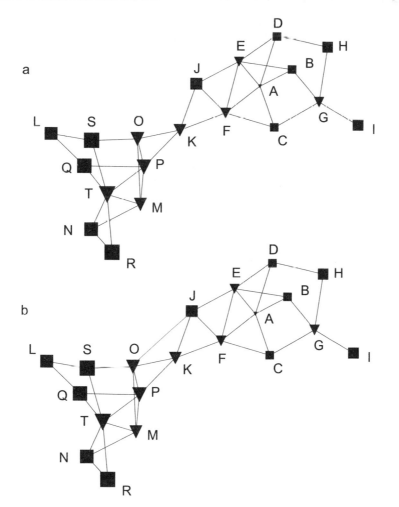

Figure 5.1. Social network of a fictitious population of *C. perfectus* (a), and the same network with one extra edge between O and J (b).

the predictor is betweenness, a measure of network centrality, shows that there are potential costs to occupying the central ground in the social structure.

Well, perhaps, but this is time to sound the first note of caution in this chapter. There will be a few of them in what follows, but we hope you will see these as useful rather than annoying. We are confident that some of the "ifs and buts" we present will, in time, be resolved, and standard procedures for statistical analysis of networks will emerge, and that they will be available at the click of a button. Our aim at this stage is merely to temper any enthusiastic

TABLE 5.1.
Top row: rank-order of the node betweenness of the nodes (highest first) in figure 5.1a.
Bottom row: rank order if an extra edge is added between nodes O and J, as in figure 5.1b.
Letters in bold indicate females.

K	F	P	O	C	T	E	G	J	A	D	M	S	Q	B	H	N	L	I	R
F	**O**	**K**	J	**P**	E	T	C	G	**M**	S	D	**A**	**B**	Q	H	N	L	I	R

tendency to over-analyze animal social networks, and also to stimulate progress in the methodological treatment of network measures.

So, back to our fictitious animal population. Our first note of caution should not be a great surprise to anyone used to "regular" statistics, and concerns the use of the node value (descriptive statistic) of one, or a small number, of animals as a basis for analysis. To illustrate how sensitive networks can be to such things, let us imagine that during our first field season we sneezed just as animal O and animal J were about to interact so we missed the edge that should have been placed between them. If we had observed this single extra interaction, our network would have been the one shown in figure 5.1b. Table 5.1 shows what the presence of this single edge does to the rank order (highest first) of the node betweenness of each animal. We see that the betweenness values of our three captured animals (K, F, and O) would have been in a completely different order, and our conclusions might have been rather different.

Of course, the solution to this alarming ability of a single edge to change the details of node measures is to look for correlations or tendencies over categories of animals, not individuals. In table 5.1, the letters in bold indicate females, and it is easy to see that females predominantly occupy the high-betweenness positions in the ranks of both the original and the "sneeze-corrected" network. Thus it is much safer to try to show that "females tend to occupy more central positions in the network" and to relate that to the parasite load, or any other biology, than to try to deduce too much from the values of individuals.

Were the data in table 5.1 derived from a conventional, attribute-based measure of each of the nodes, it would be a straightforward matter to show that "females tend to have higher values than males." A simple Mann-Whitney test (Fowler, Cohen, and Jarvis 1998) or something similar would do the job. However, in the case of a network, the node values are *not* independent of each other. It is easy to see in figure 5.1a why individual K has a high betweenness, as virtually all paths between pairs of *C. perfectus* pass through K. But these same paths all pass through either P or O, so they too will have relatively high betweenness, based on the same information.

This brings us to the main point of this chapter. Due to the relational nature of the data in a network, one of the most productive (and reliable) ways to test for statistical significance is through some kind of randomization test. Various types of randomization (or Monte Carlo) tests will be presented in this chapter, and they will be used again in chapters 6 and 7, so it is time to introduce them.

5.2 ESSENTIALS OF RANDOMIZATION METHODS

The essential idea of a randomization test is to use a particular measure derived from a data set as a test statistic. In this chapter, the measure is likely to be a degree, or a clustering coefficient, or some other node-based measure—we will refer to it here as A, whatever it is. The aim is to determine whether the value of A could have arisen by chance. To do this, a computer is used to shuffle the real data in some way to produce random versions of the data. The idea is that this shuffling will scramble any biological (or other) structure in the data that contributed to A. (In this book we will tend to refer to a randomization test using a constrained shuffle, in which some of the structure of the data is scrambled but some preserved, as a Monte Carlo randomization test. The art in these tests is to know what to scramble, and what to preserve, but we'll leave this nicety until later. We won't be too fussy about the different names for various tests involving randomization of data; we hope that it will be fairly evident from the text what we are scrambling and what we are leaving alone.)

Whatever the method used to generate it, each randomized data set is analyzed in the same way as the real set to produce a randomized value of A. This process is repeated many (typically 1,000) times, yielding a distribution of values of A that would be expected if indeed A is merely a reflection of the properties we have preserved and not those we have randomized. Then if the true value of A is in the outer 5 percent of the distribution, we deduce that its value is statistically different from our randomizations, and so is influenced by some of the properties that we've scrambled. Manly (1997) provides an excellent account of randomization techniques, including Monte Carlo and other related methods such as the bootstrap (Good 2000).

In this and subsequent chapters we will make a lot of use of randomization and Monte Carlo tests, some of which become rather involved, particularly when we are trying to unravel the structure of networks constructed via the "gambit of the group." But first let's apply the idea in a rather simple way, by returning to our fictitious example of the C. perfectus networks in figure 5.1. We want to show that in table 5.1 it is really true that females rather than males tend to occupy positions to the left (that is, they have higher node betweenness). Our test statistic (A) is some measure of the betweenness of the females. We should probably stop now to consider whether this should be the mean, or the median, value of betweenness among the females, or whether we should employ ranks rather than values; but instead we will plow on, and plump for our test statistic A being the median betweenness of the nine females.

For the network in figure 5.1a, $A = 25.9$. Now we have to decide what to randomize. In this case, we want to show that, given the network in figure 5.1, there is a tendency for the nodes (animals) with high betweenness to be female. To see if this is true, all we have to do is scramble which node is associated with which animal. That is, we randomise the *node labels*. This is a relatively straightforward procedure to perform, and can be achieved, for example, using the "resample" capability in the POPTOOLS add-in to the Excel program—see box 1.1 in chapter 1 for more details. In 1,000 randomizations of the node labels in figure 5.1, POPTOOLS found that the median betweenness of the nine nodes randomly labeled as female to be greater than 25.9 on 14 occasions. Thus we deduce that $A = 25.9$ is a statistically significant value (with a P-value of $14/1000 = 0.014$) and thus that females do tend to have higher betweenness than males. To complete the story of this example, the median betweenness of females in the "sneeze-corrected" network (figure 5.1b), with the single edge added between nodes O and J, is 36.3, a value exceeded by 10 of 1,000 node-label randomizations, so again the betweenness of females is found to be significantly higher (with $P = 10/1000 = 0.01$) than that of males. Of course, we should not get carried away here, even when analyzing fictitiously perfect data. Without replication, we cannot say anything about whether it is a general truth in *C. perfectus* that females have higher betweenness than males, although it does appear to be a real effect in this particular network.

The ideas contained in this simple example have been used to analyze real animal social networks. For example, Lusseau (2007) used a social network of bottlenose dolphins (*T. truncatus*) to establish that males and females exhibiting particular (and rare) nonvocal signals tend to have significantly larger node betweenness than other males and females. He used this node-based analysis to hypothesize about the role of social position in decision-making in this species.

What should we learn from this fictitious example? Two things: first that some node measures are very sensitive to small sampling errors, so it is wise not to build too many conclusions on the relative values of these measures for one or a few nodes, however confident you might be in your sampling protocol. Second, if we want to test the significance of observed patterns in social networks, there are issues with using standard statistical techniques because different individuals in a network are not truly independent from one another. One way around this is to compare our observed network measures with those generated by a series of randomizations of the data.

5.3 CONTROLLING FOR THE SAMPLING PROTOCOL

Of course, one of the reasons we chose our fictitious *C. perfectus* network to kick off this chapter was because it was about as easy an example as one could hope for. In any real situation, it is extremely likely that the sampling protocol will have a (sometimes profound) effect on the structure of a social network, even without "sampling errors" such as missed interactions or misidentified individuals, and we must control for this before we infer any biological meaning. For example, the number of times an individual is observed (or recaptured) will greatly affect its location in the network and subsequent network measurements, irrespective of its behavior during those observation periods. If one category of animal is ten times more visible than any other, does the fact that its network degree is twice as big as others reflect anything of interest? This is where randomization-based tests begin to come into their own, in the form of Monte Carlo simulations, in which the important features of the data, such as the number of times an individual is observed, are preserved in each randomization. These randomizations are essentially "null models" of social interactions or associations.

Issues related to variance in the number of times an individual is seen occur in virtually all observations of wild animals that we can think of. However, there is an even more important constraint on network structure if, as has often been the case so far in applications of network theory to animal populations, we have used the "gambit of the group" to construct networks. Then the group-size distribution and the fact that each observed group contributes a perfectly connected cluster to the network will greatly affect node-based measures. We have already discussed in chapter 4 the fact that the clustering coefficient is always likely to be high in a network constructed via the gambit of the group, as perfectly connected groups contain many triangles.

Due to the prevalence of group-based animal social networks in the recent literature, and our expectation that there are many existing data sets that could be analyzed using the gambit of the group, our emphasis for most of the rest of this chapter will be on testing node-based measures in such networks.

Two Methods for Randomizing Group-Based Data

The issue, then, is to see whether simple node-based measures (L, C, k, B, etc.) and their distributions over different classes such as sex can be reliable indicators of something biologically interesting in a group-based animal social network, and are not merely artifacts of the sampling strategy. The best approach we have at the moment is to randomize the data in a way that preserves the group structure and recapture frequency, but scrambles the tendency of pairs of animals to be seen together in a group. There are two methods that achieve this, each developed for a slightly different scenario. In both cases, the essential point is that we will randomize over edges, not node labels. That is, we will redistribute the pair-wise relations in some way. As with the randomization of node labels in section 5.2, the basic premise is that a comparison of the observed social structure with one that is obtained if all individuals in our group or population interacted randomly with one another provides us with a way of testing if our observed data could have been brought about by chance (Manly 1997). To test our hypothesis we need to choose a test statistic (one of our node measures, in this chapter) that reflects the pattern of interest. We then simulate random interactions between individuals on the computer. For each simulation we calculate the test statistic and rank the values obtained. Finally we compare the observed value of the test statistic with the generated ones (from the simulation). For a two-tailed test, the observed value must be in the lower or upper 2.5 percent of the distribution of simulated values for us to reject the null hypothesis that the observed pattern was generated by chance. In contrast, for a one-tailed test (i.e., we have a directional prediction that the observed value should lie in one of the two tails of the distribution), the null hypothesis is rejected if the observed value falls with 5 percent of the a priori predicted tail of the distribution. As an aside, we should remind ourselves that detecting nonrandomness in our observed data is usually just the starting point for a more detailed investigation of the observed pattern(s), and not an end in itself.

Initially one may think that there is only one way of carrying out a randomization of one's own observed data set. Actually there are many ways of going about this, and the choice of the appropriate randomization technique has important implications for the conclusions that can be reached regarding the observed data. Therefore a certain amount of care must be taken in deciding how to generate the randomized networks (see Whitehead and Dufault 1999; Whitehead, Bejder, and Ottensmeyer 2005) and in particular what features of the data set to conserve during the randomization process. The most obvious randomization strategy is simply to rewire the observed network, lifting existing edges and placing each between two randomly chosen nodes. This generates networks in which the number of pair-wise associations is preserved, but little else, certainly not the group-size distribution, which is likely to have a large effect on network measures, as we shall see.

Before we go any further, it will perhaps help to clarify the difference between the number of pair-wise relations represented by a network (P), and the number of edges (E). In an unweighted network, $E = P$. Our simple rewiring randomization strategy in this case will produce what we referred to in section 4.6 as an Erdös-Rényi random network. By ensuring that the same edge is chosen once at most, the randomized network will have the same value of E (and P) as the original, as noted above. However, if we have a network collated via group membership, some pairs of animals may be found together more than once. This is represented not by having several edges between their nodes, but by a single edge carrying a weight equal to the number of co-occurrences of that pair in the data. Then E is the number of distinct edges, which is always less than or equal to the number of pair-wise associations P. Now we have a choice: we randomize either the E distinct edges or the P associations. If the former, we are randomizing the network edges. If the latter, we are randomizing the associations. We will concentrate on this latter option, since it takes us back to the original data set. In this case it makes no sense to restrict the number of times two nodes can be connected in the randomized network, which thus has weighted edges, just like the original. Now P is preserved by the randomization, but E need not be. We will refer to the scheme of re-wiring P associations as the "obvious randomization" or "OR" later in the chapter.

The two randomization methods we will present are designed to do better than the simple OR rewiring scheme outlined above, in that they each conserve two of the features of a group-based data set that are found to be particularly important, namely the group sizes and the recapture frequencies (the number of times each animal is observed). To achieve this, they randomize not at the level of a pair-wise association but by scrambling the membership of groups, then constructing a randomized network by joining together all animals in the randomized groups.

The first method is based on repetitions of a simple process of randomly choosing two animals in different groups and swapping them. Swapping continues until the membership of the groups is scrambled. As far as we are aware, this method was developed by Manly (1995) to solve an ecological problem, and picked up by Bejder, Fletcher, and Bräger (1998) in the context of testing for nonrandom social associations in group-based data. It was coined the "MBFB method" by Whitehead (1999) in honor of the four authors of these papers. The method has generally been used to randomize group-based relational data in protocols where one group is observed at a time. It was extended by Whitehead (1999) to consider more constrained randomization where, for example, swaps only occur within groups that are a subset of the whole data set, and again by Whitehead, Bejder, and Ottensmeyer (2005) to constrain swaps within specified classes or categories of animal. These randomizations can be performed within Whitehead's Socprog computer program (see box 1.1 in chapter 1). Similar phenotype-restricted swaps were also used by Ruxton and coworkers (Hoare et al. 2000) to tease apart species and body-length effects from parasite-load effects in the composition of shoals of small freshwater fish.

The second method, developed by James and coworkers and first presented in Ward et al. (2002), was designed to randomize group membership within data sets gathered from a series of censuses of group membership. In a single census, the membership of all observable groups is determined at once (or nearly so, depending on the census protocol). Then each animal can belong to at most one group in each census. This second method does not start from the observed group memberships and perform pair-wise swaps. Instead it simply takes all of the animals from a given census, and re-allocates them at random to groups of the same sizes as those observed. We will call this scheme the "GR" or "group-based randomization." Any subsequent re-assortment of the groups, to reflect phenotypic or other heterogeneity, may in principle be

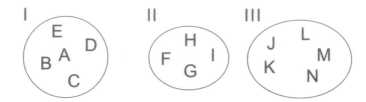

Figure 5.2. Three groups containing 14 individually identified animals, labeled A–N.

achieved by subsequent pair-wise swapping of animals, in the spirit of the modified MBFB method.

To illustrate the basis of these two randomization techniques, consider the simple cartoon in figure 5.2, which depicts fourteen animals (labeled A–N) in three groups (I, II, and III). Method 1 might swap A with F, then H with L, and so on. Assuming the three groups comprise the whole census, method 2 would deal with all fourteen animals in one shot by placing them randomly into groups of size 5, 4, and 5. In each method a certain amount of computational "bookkeeping" is required to make sure the randomizations are truly random and unbiased, but they amount to the same thing. Each offers a slightly more subtle approach to randomization that we have so far dealt with, because the randomization is constrained by features of our choosing. This difference is usually expressed by referring to tests based on procedures like this one as Monte Carlo tests rather than randomization tests (Manly 1997). We might equally choose to refer to such randomizations as "null models" of network formation.

If all of this is becoming too methodological for your palate, perhaps it is time to pause and reflect on why it should matter what randomization test we use. Surely random is random? Well, no. From a practical perspective, we will see in section 5.5 that the choice of what to conserve and what to randomize can make a tangible difference to node values in a "null model" of network formation. By choosing to constrain the group sizes and recapture frequencies, we are constructing a null model of what we think the most important factors are that should be controlled for before we search the data for meaningful biology. From a biological perspective, detecting nonrandomness is a fairly trivial exercise in itself, and we usually want to know more than this about our study system. For instance, we may want to tease apart different factors by controlling for the influence of one while randomizing our data set with respect to the other (Hoare et al. 2000; Whitehead, Bejder, and Ottensmeyer 2005). Which biological features of the observed network need to be preserved and which ones are subject to randomization depends on the question that the test is meant to answer. Let's say we are interested in the composition of multi-species groups where individuals also vary in body length. In this case we might keep the group size distribution as it was observed and carry out pair-wise random-izations of individuals between groups that are of similar body length but differ by species. This test would then allow us to determine whether in the absence of body length variation between groups, there is any degree of species assort-edness in our system. The choice of the appropriate randomization technique has important implications for the conclusions that can be reached regarding the observed data. If we constrain either too little or too much in our data set, we may not have sufficient power to detect the things of interest.

5.4 FILTERING EDGES

Before we can turn our randomization methods to the job of trying to extract some biology from a group-derived network, there is one more methodological hurdle we must jump. We have already mentioned several times that using the "gambit of the group" to construct networks introduces many edges that prob-ably do not represent potential "relationships" (in the sense of Hinde [1976]). These "unwanted" edges will nonetheless contribute to each of our node mea-sures, and potentially cloud any useful information these measures might con-vey. Now is the time to do something about them by filtering out some of the edges we deem less likely to represent meaningful social interactions. In many ways this might seem an undesirable step, as we will be throwing away infor-mation. In addition, the "importance of weak connections" (Granovetter 1974) is well known in studies of the spread of information, for example. However, when taking the gambit of the group, we have no way of distinguishing arbi-trary connections between group members from weak but potentially important edges. Thus it seems reasonable that we should accept that the safest patterns to look for in group-derived networks are those corresponding to strong or core connections among pairs or groups of animals.

Given this position, we might expect on the one hand that in terms of filter-ing, "the more the merrier." If filtering distills our choice of edges to those that really count, let's be very strict and remove as much as possible. On the other hand, we don't want to filter so hard that there is nothing left to analyze! Again, as with randomization methods, we have to stress that there is not yet a foolproof and generally accepted way to go about filtering group-derived net-works, though we hope that this situation will improve in time. In this section we will present two alternative methods, each with pros and cons, which at the very least serve as a vehicle for discussing this issue.

The first method, used in a series of papers analyzing dolphin social networks (Lusseau 2003; Lusseau et al. 2003; Lusseau and Newman 2004; Lusseau et al. 2006), uses the SocPROG program to randomize the observed network using the MBFB method outlined in section 5.3. In these analyses an observation threshold (see section 3.4) is first applied to remove animals seen only rarely.

The randomizations are then used to construct the dolphin network by computing a "dyadic P-value" for each edge using the "half-weight index" (HWI), an association index (see box 3.3 in chapter 3) of that edge, as a test statistic. An edge is placed in the dolphin social network if its HWI falls in the upper 2.5 percent (Lusseau et al. 2003) or 5 percent (Lusseau 2003) of the randomized HWI values (generated by the MBFB method) for that edge. All the other candidate edges are filtered out.

There are considerable attractions to this approach. The use of an association index certainly helps to ameliorate the effect of many sampling biases, including the creation of many unwanted edges, and the use of randomization is a good idea when dealing with relational data sets. SOCPROG is freely available, so all of the nasty computing details are largely taken care of. What is more, if the "difficult" statistical issues can be dealt with at the stage of network construction, the way is paved for much more straightforward analysis on the subsequent network. This method appears then, on the face of it, to pass our first criterion outlined at the start of this chapter for an easy life; namely that it seems to place the choice of edges on a sound statistical footing, enabling us to create "the" social network for our system, which we subsequently analyze. On the basis of this approach, a considerable amount of interesting biology has been explored (e.g., Lusseau and Newman 2004; Lusseau 2007; Lusseau et al. 2006) that has pioneered the networks approach in animal systems.

Unfortunately, the statistical basis for one step in this process is not robust. Whitehead, Bejder, and Ottensmeyer (2005) pointed out that the method of constructing a network by calculating dyadic P-values is "conceptually invalid." They say that "dyadic P-values should not be used as a measure of the strength of the relationship between two animals. This is the purpose of the association index itself." This is essentially saying that any P-value derives from two factors, the sample size and effect strength (the HWI in this case), and, unless you are very sure that you can control for these, you have no idea how much of the P-value is due to one factor or the other. So while it might be safe to use a dyadic P-value to determine that the presence of one particular edge is significant, it is not safe to compare P-values from many edges and use this comparison as the basis of acceptance or rejection of an edge in the network.

These arguments may seem a trifle arcane to most of us, but they do matter. Our interpretation of the situation at the moment is that this avenue is an interesting one to explore further, and still has many features to recommend it. The use of an appropriate association index (AI), rather than a simple "association strength," to measure the relative weight of edges is a sensible idea, but there is still a considerable degree of arbitrariness in how to choose both an observation threshold and an AI threshold, as dyadic P-values are not to be trusted. It is interesting to note that Lusseau et al. (2006) acknowledge a degree of arbitrariness in the method when they choose to compare their network with one where the mean HWI is used as a basis for including or excluding edges.

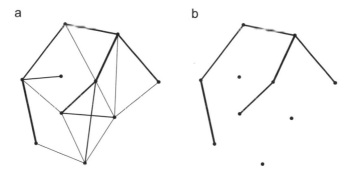

Figure 5.3. A simple weighted network (a) with edge weights represented by line thickness. Only those edges with a weight greater than or equal to some threshold *T* are included in the filtered network (b). Edges in the filtered network are all assumed to have unit weight.

Happily, that particular analysis yields very similar results for the two filtered networks. Of course, none of the issues with dyadic P-values negates the use of the MBFB method of randomization. More recently, Lusseau, Whitehead, and Gero (2008) have developed new and interesting methods that use the MBFB randomization technique in a more robust way.

In analyzing our freshwater fish networks, and others, the authors have tended to adopt not a "two-stage" process (construct the network, then analyze it) but always to compare our observed networks directly with those constructed via a randomization of the data. As it happens, we have always used the second of the randomization approaches outlined in section 5.3, but this is a detail. We have also tended to use just the "association strength" (AS)—the number of times (censuses) two animals were seen together—as the measure of edge weight, and used different (integer) threshold values of AS as a simple means of filtering the edges. An "AS3" filter keeps only edges representing animals that were seen together three or more times. As with Lusseau's methods, any edge that survives the filter is subsequently treated as having weight 1; those that are chopped are given weight zero, of course. See figure 5.3 for a simple illustration of this.

The advantages of this approach are that an AS filter is very simple to apply, and more importantly, that the emphasis is on randomizing the biological data, not the network. An edge between two chosen individuals may or may not appear in a particular randomization, but this is addressed through our statistical tests on network measures. Disadvantages include the fact that AS values are crude determinants of a cutoff (though any of the association indices could equally well be used) and, at this stage at least, rather arbitrary. Perhaps more importantly, the number of distinct edges *E* is not conserved by this "one-stage" randomization procedure, and this will certainly have an effect on the network measures obtained. However, all of this can be monitored, as we will now reveal via a case study.

5.5 AS-FILTERED RED DEER NETWORKS

In this study we will use a single data set to illustrate some of the issues surrounding the choices of randomization (or Monte Carlo test, if you prefer) and strength of the edge filter when analyzing animal social networks constructed via the "gambit of the group." The analysis is not intended as a recipe for others to follow, but rather as a means to highlight the importance of taking these two issues into account, and to guide the reader through the systematic breaking down of a social network in the hope of providing a better understanding of why they look the way they do. The edge filtering will be based on choosing an AS cutoff, as described at the end of the previous section. A subsidiary aim is to start a discussion on an appropriate rule of thumb for deciding how hard to prune a network (by edge filtering) in order to reveal the biology of interest.

Our model data set is but a small slice of the long-term investigation of the red deer (*C. elaphus*) population on the island of Rum (Clutton-Brock, Guinness, and Albon 1982), kindly made available to us by Tim Clutton-Brock. One of the pieces of information gathered in this study has been the identity of members of groups of deer during regular visual censuses. Our illustrative data set is taken from the 26 censuses of group membership between January and May 1990. There are 342 deer (212 female, 130 male) in the "starting network," in which all edges are included. Our first series of tests will look at what happens to the mean values of degree (k), node-based path length (L), clustering coefficient (C), and node betweenness centrality (B) at various values of AS filter cutoff. (We should note as an aside that were we to choose a single index to characterize the overall centrality of a network, we might be tempted to base that measure on the spread of the node values, not their mean, as we have used here for convenience. Such measures of spread, often a variance, are termed network "centralization" [Wasserman and Faust 1994].) In the second set of tests, we explore the hypothesis that since red deer are a classic example of a species exhibiting sex differences in social behavior (Clutton-Brock, Guinness, and Albon 1982), at least some of our node-based network measures ought to show sex-related differences. Again the extent to which this is true in our network is investigated as a function of AS filter cutoff.

To test for the statistical significance of our results, we will compare our results with two separate Monte Carlo randomizations. The first is the "obvious" randomization (OR) introduced in section 5.3, in which we take all P pair-wise associations from the original data and place each of them between two randomly chosen nodes. The second is the group-based census randomization (GR), also described in section 5.3, which preserves the group sizes and observation frequencies. Randomized networks are constructed in exactly the same way as the real one; a "weighted association matrix" $\mathbf{W}$ (section 4.7) is constructed, whose entry W_{ij} denotes the number of times deer i and deer j were seen in the same group (either for real or within a randomization). Then the unweighted association matrix $\mathbf{M}$ is built. If the value of the AS filter is T, say, then $M_{ij} = 1$ if W_{ij} is T or greater, and zero otherwise (see figure 5.3). $\mathbf{M}$ is then analyzed to produce the required node values.

To begin, we should follow our own advice of chapter 3 and visualize our network. Figure 5.4a shows a small fraction, containing 46 nodes, of the raw network, with an edge between two nodes if those two deer were ever seen in the same group in any census. In the language of this chapter, this is part of the "AS1 network." (The entire AS1 network, for all 342 deer, is an impenetrable tangle of nodes and edges, so we have shown just a small part of it for the purposes of illustration.) For comparison, Figure 5.4b shows the AS6 network for the same 46 nodes, in which edges only appear if the two animals were seen in the same group during at least 6 of the 26 censuses. Nodes that become isolated through the filtering process have been removed.

The first thing to notice about the AS1 network is that, visually at least, it is a mess (though nowhere near as bad as it looks with all 342 nodes included). It is difficult to see how this graph could help us begin to formulate tests of the structure of this network—there are simply too many edges, introduced via the gambit of the group. The AS6 network, by contrast, both for the subset of nodes we have chosen and for the full set of nodes (not shown), looks more appealing to the eye, containing discernable structure in terms of weakly inter-connected clumps of nodes (see chapter 6) that immediately grab our attention and might induce us to take a more careful look. However, at this stage we have absolutely no justification for choosing to analyze the AS6 network other than that it "looks good."

Mean Node Values

Figure 5.5 shows plots of mean values, over all 342 nodes, of the degree (k) and clustering coefficient (C) as the AS cutoff value T is increased from 1 to 7 in our full deer network. In this and subsequent figures in this section, values for the observed network are denoted with a diamond ($\blacklozenge$), a square ($\blacksquare$) denotes the mean value derived from 100 randomizations of type OR, and a triangle ($\blacktriangle$) denotes the mean value from 100 randomizations of type GR. (Note that 100 randomizations are not as many as we should use were we truly trying to say anything here about the red deer population on Rum. They are sufficient, how-ever, to present the salient issues in this section. For the sake of argument, we will declare the real node measure to be statistically different from a random-ization if it is in the bottom or top three of the randomized values.)

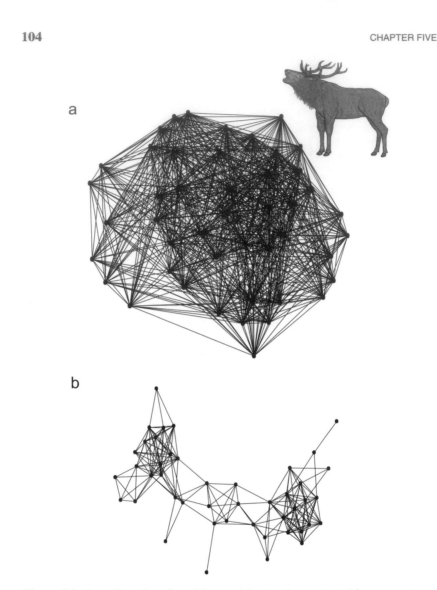

Figure 5.4. A small portion of a red deer social network constructed from group-based data. (a) is the unfiltered network, (b) is the same network filtered with a minimum association strength (AS) of six.

What can we learn from these plots? Let's take the mean degree first. There is not a great deal of difference between the two randomizations in this case, but each is different (significantly by the approximate criterion set out above) from the observed values, at all filter strengths. For AS1 we have $k(\text{random}) > k(\text{observed})$, but this is reversed for all higher AS cutoffs T. This is easy to

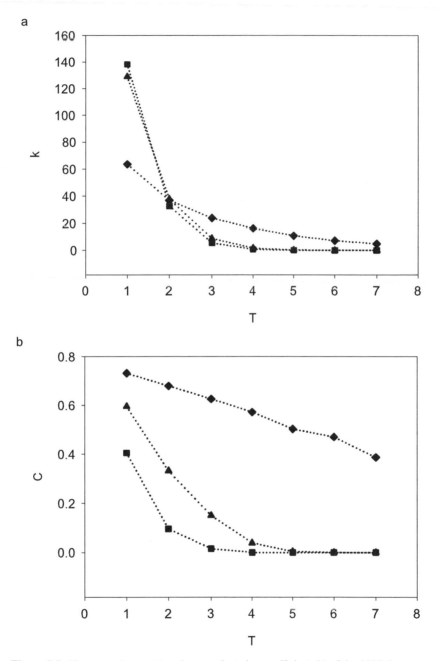

Figure 5.5. The mean degree (a) and mean clustering coefficient (b) of the 1990 deer network as a function of AS filter cutoff threshold T. Diamonds represent the real data, squares those from 100 OR randomizations, and triangles those from 100 GR randomizations.

understand in terms of the prevalence of "repeated associations." In the real network the edges are concentrated on relatively few pairs, with plenty of edge weights of 2 or more. The same number of edges is spread more evenly in the random networks, giving more edges at $T = 1$ and fewer at high T. This difference between observed and random mean degree is potentially meaningful, as it does suggest that something (be it preferred habitat use, active choice, or whatever) is causing some animals to be placed in the same group as each other more often than would be expected by chance.

Figure 5.5b shows the mean clustering coefficients from our calculations. The value for the real network falls steadily with T, indicating that as we remove more and more weak edges, we are breaking apart more and more triangles, so that those that remain are not there due to a single chance co-membership of a group. This time, the OR and GR randomizations give very different results, at least up to $T = 4$ or so. In GR we preserve group sizes, and hence the high "census-level" clustering. Nonetheless, the observed network appears to be more clustered than expected from the Monte Carlo tests. Again, all differences between the observed and randomized values are statistically significant, and therefore potentially contain interesting information about the social structure of the deer population.

Note that the two randomizations each give similar mean values of C as T increases above approximately 5 or 6. This reflects the reduction in k as we filter edges out. Eventually, as k becomes less than 2 or so, the probability of a triangle appearing (which needs 3 neighboring nodes all to have a degree of at least 2) is drastically reduced, and the clustering coefficient vanishes to zero. Beyond this point we are probably beginning to filter the network (and especially its randomized counterparts, with fewer surviving edges still) to death.

This breakup of the networks is particularly noticeable in the plots of the mean node path length L among "reachable pairs" (section 4.2), and in the betweenness B that is derived from path lengths. See figure 5.6. These measures tend first to increase with increasing T, reflecting the fact that filtering removes direct paths between nodes, so the shortest paths become more tortuous. As T increases further, the network breaks into more than one large component and the L and B values become smaller again, as they are reflecting paths within two (or possibly more) smaller components. This tends to happen more in the randomized networks than in the real networks at a given T; again this is because the former have fewer edges. (As a yardstick, a large Erdös-Rényi random graph will fall apart if its mean degree falls below 1 [see section 4.2 or, e.g., Newman 2003a]. Our randomized networks are more structured, but will still break apart into two or more large components at some stage as k is reduced.) The potential of filtering to break up our networks, and the fact that this will not occur in all randomizations, means than both L and B should be treated with slightly more caution than k or C when inferring that the observed values are statistically large or small.

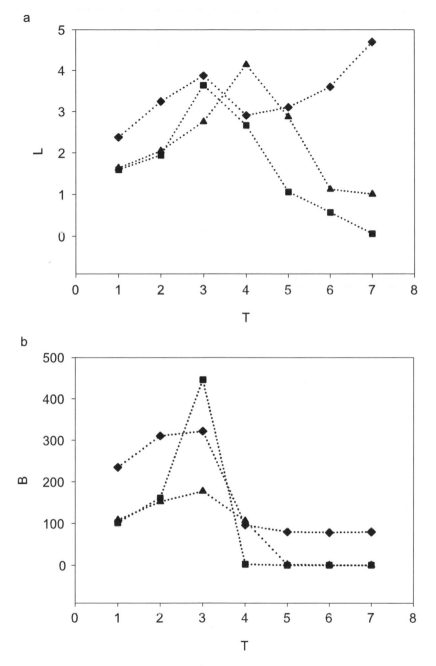

Figure 5.6. The mean node path length among reachable pairs (a), and the node betweenness (b) of the 1990 deer network as a function of AS filter cutoff threshold *T*. Diamonds represent the real data, squares those from 100 OR randomizations, and triangles those from 100 GR randomizations.

Though it would be folly to generalize our findings at this stage, we have observed qualitatively rather similar patterns in the plots of all the node values shown in figures 5.5 and 5.6 as a function of T in our networks of small freshwater fish. Indeed, Croft et al. (2004a; 2006) used the difference between k(observed) and k(GR)—though presented in a slightly different form—in their study of social networks of the Trinidadian guppy (*P. reticulata*). They found that the number of "persistent pairs" (which they defined as edges with AS = 3 or more) was greater than expected from a group-preserving (GR) randomization test. This is essentially stating that the mean degree at AS3 is significantly large. We have found the same result here for the red deer networks, at a range of values AS cutoff T. Croft et al. (2006) go on to analyze the persistent pairs by the sex of the animals, finding that it is only the female–female pairs that occur more often than predicted from the null model, and investigate the potential consequences of these preferred pairings.

Separation of Node Measures by Category?

Let us pick up this baton of sex differences in filtered networks and run with it a little. As biologists we want to explore the mechanisms underlying structure, so we might want to look, among other things, for evidence of phenotypic assortment in our node measures. As we have already said, sex segregation does occur in red deer (Clutton-Brock, Guinness, and Albon 1982), so it would seem odd if none of our node-based measures exhibit differences between the sexes. (Conversely, there is no reason to assume that *every* metric will show sex differences—it will depend on how sex segregation is expressed in the deer and how the metric probes the structure of the network. To illustrate this point with an extreme example, suppose we had a species in which every single animal had the same number of social relations, but in which females were far more cliquey than males. A social network for this species would be expected to show sex differences in individual clustering coefficients, but not in individual degrees.) In this case study we will use sex as a simple means to classify our nodes and look to see whether node measures show any evidence of being differentiated by category.

Figure 5.7a shows the frequency distribution of node-based path length L_i among "reachable pairs" for the AS3 (i.e., filtered at $T = 3$) red deer network, plotted separately for males and females. There clearly seems to be some separation between the sexes, and you are invited to think about what this might mean biologically. Our concern here is to determine whether such a separation is statistically robust. The statistical challenge is that a simple test can't tell us this because of the nonindependence of the data. We might be tempted by a node-label randomization similar to the one we used in section 5.1, but this would be taking the network structure as a given, rather than subjecting it to a test that randomizes the social associations. Figure 5.7b shows the frequency

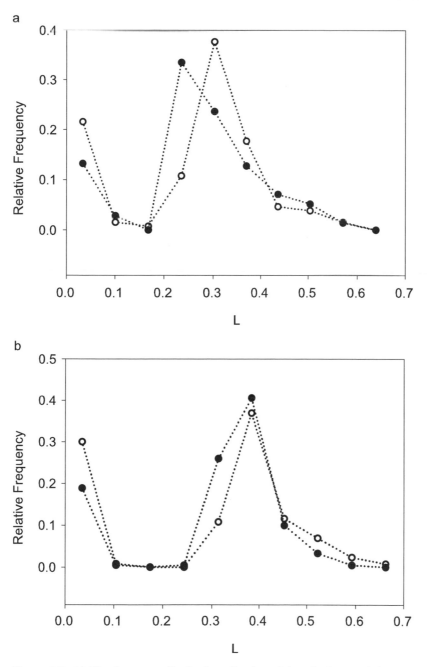

Figure 5.7. (a) The frequency distribution of node path lengths for a red deer social network filtered at AS3. Open circles are males; filled circles are females. (b) The frequency distribution taken from a single GR-type randomization of the same data.

distribution of L_i from a single randomization of group membership in each census, using the GR algorithm outlined above. There appears to be a certain amount of separation by sex here too, so we must proceed with caution. We will again employ both the "obvious randomization" (OR) and the GR randomization that preserves all group sizes and recapture frequencies. To test whether the node values we calculate are more (or less) separated than we would expect by chance, we use a ranks test based on a test statistic u and a P-value—see box 5.1 for details.

Box 5.1

Testing Node-Based Measures for Six Differences

We have n_1 females and n_2 males. Let the node measure under consideration be A. Then for a given network (the observed one or a randomization) we rank all $n = n_1 + n_2$ values of A (smallest first) and find the sum of the ranks among females, R_1. In a "conventional" Mann-Whitney test (Siegel and Castellan 1988), the statistic U is calculated as $U_1 = n_1 n_2 + \frac{1}{2} n_1(n_1 + 1) - R_1$ or $U_2 = n_1 n_2 - U_1$, whichever is smaller. By choosing the lesser of these two values, we lose information on which category had the lower ranks. In the current situation we wish to know whether females or males have lower ranks, so our test statistic u should be based on either U_1 or U_2. We arbitrarily choose U_1, and scale it by $n_1 n_2$ (which happens to be the sum of U_1 and U_2). Thus

$$u = \frac{U_1}{n_1 n_2}.$$

The value of u is always between 0 and 1. If females occupy the lowest n_1 ranks of A, $u = 1$, and if males occupy the lowest n_2 ranks, $u = 0$. If there is perfect mixing between the ranks of males and females, $u = 0.5$. To test whether an observed value of u is statistically significant, it will be compared with 100 values drawn from a Monte Carlo simulation of the data to produce a P-value. If $u(\text{observed}) < 0.5$, then all $u(\text{random}) > u(\text{observed})$ support the hypothesis that $u(\text{observed})$ could not have arisen by chance. If $u(\text{observed}) > 0.5$, then all $u(\text{random}) < u(\text{observed})$ support the hypothesis. Thus a P-value of 0.02 would indicate that the observed node measures A are more sex-separated than we would expect by chance; a P-value of 0.98 indicates that the values of A are statistically less sex-separated than we would expect by chance.

Figure 5.8 shows the values of the test statistic u for the mean degree k and clustering coefficient C as the AS filter cutoff T is varied between 1 and 7 for our deer network. The u-values shown for the OR and GR networks are the mean values over 100 replicates. Similar results are found for the mean path length L and node betweenness B. The first thing to notice is that for the real network, each of the measures k and C becomes more differentiated by sex (u gets further from 0.5) with increasing levels of edge filtering. This adds weight to our earlier hunch that filtering our "gambit of the group" network may sharpen the information it contains. The u-value for the OR randomization is almost exactly equal to 0.5, which indicates perfect mixing of the node values among males and females (see box 5.1). This is what we might naively expect from a "random network." However, the key thing to note is that the randomizations (GR) that preserve group sizes and recapture frequencies, but scramble which deer are in which group in a given census, themselves yield node values that are rather sex-separated. So (at last, you might be thinking) we can see the importance of our null model randomizations constraining at least some of the features of the original group-based data structure.

To test whether the sex separation of node values is statistically significant, we must compare the observed and randomized u-values, following the procedure outlined in box 5.1. Our naive OR randomization would lead us to believe that all four of our node measures (k, L, C, and B) are significantly more sex-separated than expected by chance, at all values of T. The GR randomization is more discerning, as can be seen in table 5.2. The approximate P-values are based on 100 GR randomizations. $P = 0$ indicates that all 100 randomizations were less sex-separated than the real network; $P = 1$ that all 100 were more sex-separated.

Table 5.2 shows that, at least for this data set, the clustering coefficient is a good discriminator between the sexes. The two centrality measures, k and B, also indicate a significant separation by sex, but only once a fairly stringent edge filter has been applied. The final measure, L, shows a similar pattern in its P-values, save for a blip at $T = 5$, where the breakup of the randomized networks into more than one large component again confounds the analysis. It is particularly interesting to note that, apart from C, all these measures would suggest that the unfiltered deer network ($T = 1$) is *less* separated by sex than expected by chance, but that a "core" network, in which the edges indicate many repeated group co-occurrences of animals, has node measures that are significantly *more* sex-separated than expected by chance. Of course, we don't know, without further study, which of these results is true, but we are tempted to suggest that the latter is the case. Then our conclusion must be that the analysis of unfiltered group-derived networks can give misleading results, and that careful filtering may be an essential step in extracting useful results from such networks.

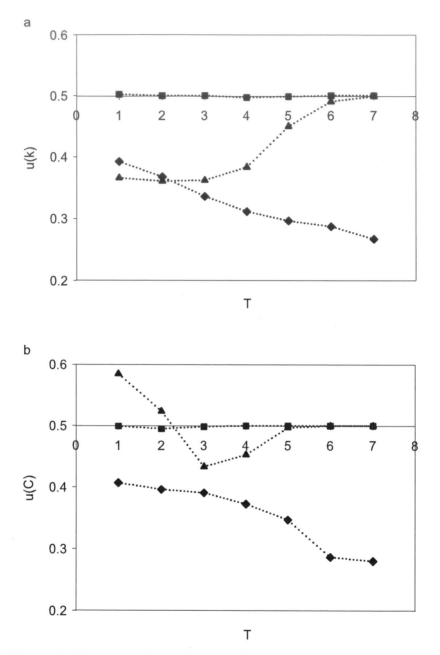

Figure 5.8. The test statistic *u* calculated for the mean degree (a) and mean clustering coefficient (b) of the 1990 deer network as a function of AS filter cutoff threshold *T*. Diamonds represent the real data, squares those from 100 OR randomizations, and triangles those from 100 GR randomizations.

TABLE 5.2.

Approximate P-values for the hypothesis in our 1990 deer network that the sex-separation in the distribution of each node measure could have arisen by chance. The test statistic is u calculated for the 1990 deer network as a function of AS filter cutoff T, and for 100 GR randomizations of the data set.

	P-value			
T	on $u(k)$	on $u(L)$	on $u(C)$	on $u(B)$
1	0.91	1	0	1
2	0.7	0.55	0	1
3	0.05	0.12	0.04	0.99
4	0	0	0	0.47
5	0	0.95	0	0
6	0	0	0	0
7	0	0	0	0

A Rule of Thumb for AS Filtering?

So if some sort of filtering is necessary, how much? At the risk of deducing too much from a single case study, it may be worth seeing whether all this effort can help us derive a rule of thumb for what level of AS filtering should be used for a given network based on the gambit of the group. Unfortunately, we cannot offer a cast-iron case at present, but a couple of possible solutions spring to mind that may warrant further exploration. There is at least some evidence in the results we have presented to suggest that AS6 or thereabouts might be a well-filtered version of this particular deer network. The mean node values and their distributions over sex categories are all "well behaved" by then, yet we appear not to have squeezed all the edges from the network.

If we accept this vague argument, how might we arrive at $T \approx 6$ as a proposed cutoff? Let's stick our necks out a little and suggest a couple of approaches. The first is based on the observation that the mean association strength (the mean number of times pairs of animals were seen together) in the raw deer network is 2.8. So we could try "filter at about twice the mean AS" as a possible guide. Of course, we could filter at some centile of the AS distribution, but all this is entirely arbitrary. As a second possible approach, we could note that our analysis of the sex-separation of node values is suggesting that the harder we filter, the "better" we do. Our earlier analysis suggests that if we keep filtering, eventually there are too few edges left to give us any test power. This is likely to have happened once the mean degree is of order 1 or so. Thus "filter until the mean degree is a little bigger than 1" might prove a useful adage—certainly a visual exploration of group-based networks often starts to reveal interesting structure once the AS filter is somewhere close to breaking up the "giant component" of the network. This latter approach is basically suggesting that there is a rough optimum between filtering out unwanted edges, ghosts of taking the gambit of the group, and losing all the information. Of course, all this is highly conjectural, but it will be interesting to see whether these or other rules of thumb will emerge that will enable us all to have a little more confidence about which network to analyze. As an aside, we note that Palla et al. (2005) come to a similar conclusion in a different context; that the interesting structure in networks might best be seen once they are filtered close to the point at which they would break up into smaller components.

5.6 OTHER APPROACHES

An alternative approach that might seem attractive after all this analysis would be to avoid filtering altogether by using weighted network measures (section 4.7) rather than the unweighted versions we have used throughout this chapter. Surely if the edges are weighted (by association strength or an appropriate association index, for example), then measures that take into account that some edges are more weighted than others will overcome the problems with the apparently arbitrary choice of how much to filter, and at what stage, right? Well, we don't yet know the answer to this question. On the face of it, weighted measures might well cut through some of the Gordian knots we have presented, and we feel this is a sensible avenue to be explored. Certainly it feels as if by filtering we are throwing away a lot of the information we have gathered. At the moment though, there is no consensus on which weighted measures might be most useful, and even if there were, it would seem that a data set would have to have a fair range of AS (say) to avoid all of the issues arising from the gambit of the group. We note that the use weighted measures is advocated in a recent paper by Lusseau, Whitehead, and Gero (2008).

A second approach is to base analysis not on node-based measures, but to look for techniques that tell us something about the structure of the network as a whole, as it is revealed through its association matrix. For example, we might use a Mantel test (Mantel 1967) to compare the overall structure of a network with a "hypothesis matrix" (Hemelrijk 1990a) representing a known pair-wise relation within the population that might be driving network structure. Alternatively, we might tap into some of the developments in the social sciences, where various models have been developed that try to capture the essential structure of a network in terms of a relatively small number of parameters. Each of these approaches may have a great deal to offer, especially if we are confident that our network is a reliable representation of the social interactions at hand. Both methods are also candidates for comparing two or

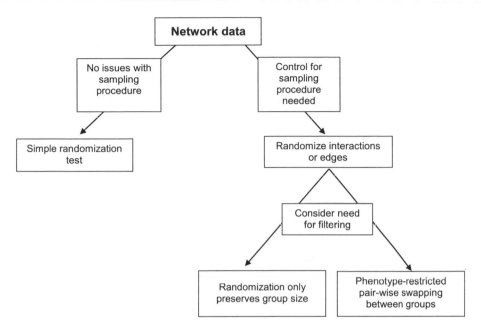

Figure 5.9. A guide to how to analyze node-based network measures of animal social structure.

more empirical networks, so rather than dive into the methodologies here, we will save a discussion of each of them until chapter 7.

5.7 SUMMARY

The main message from this chapter should perhaps be that you need to make a number of decisions before analyzing your network data. As often as not, the decisions boil down to making a choice of the appropriate randomization test to use given the structure of your data and the questions you are trying to address. The flow chart in figure 5.9 is aimed at helping you with this, though it should not be read as anything other than a guide.

The first issue is to decide whether you believe your network is likely to be strongly affected by the sampling protocol. If not, life is relatively simple, as you can follow the left side of figure 5.9. You should be wary of comparing the numerical node values of individuals, but tests on mean or median node values can be executed with a simple randomization of your node labels, as discussed in sections 5.1 and 5.2. In other circumstances, different randomizations might be used. For example, a random rewiring of the observed edges might be appropriate; there are even techniques (Newman, Strogatz, and Watts 2001) that enable you to constrain the degree distribution of each randomized network should that be the crucial element needed to tease out the biology you are interested in. Finally, you might care to analyze your network by some of the alternative schemes mentioned in section 5.6 and explored further in chapter 7.

If, like most of us at this stage, you have taken the "gambit of the group" when constructing your network, and got the benefit of all those lovely connections between animals that you can't be sure were interacting, this is payback time—you need to think a little more about the extent to which your node values and their means and distributions are affected by the structure of your data set. The methods we have set out in sections 5.3–5.5 control for the sampling protocol by randomizing not the node labels or the edges directly, but the group membership. They are thus built upon simple null models of social association. As we illustrated using a red deer social network (section 5.5), conserving the group sizes and recapture frequencies appears to be a minimum requirement of a null model. Further refinement of the null model, by constraining phenotypic variation within groups for example, might also be considered (see Hoare et al. [2000], and Whitehead, Bejder, and Ottensmeyer [2005] for examples of this type of analysis). We also showed in section 5.5 that filtering a group-derived network can have a considerable effect on any conclusions we might want to draw, so some thought should be given to this issue as null models are applied. More work is needed before we can with any confidence produce a recipe for how best to deal with group-derived networks. In particular, it would be very useful to have a reliable rule of thumb to help us decide how to apply a filter to nodes or edges or both. Until then, we should perhaps be relatively conservative in what we try to claim our network is revealing.

6

Searching for Substructures

Up to this point the emphasis of our analysis of animal social networks has been to identify some individual-based measure (such as the degree k_i or cluster coefficient C_i, for example) and to see whether the distribution of this measure is biologically revealing. In this chapter we move on to take a more global look at the structure of a network, and to search for evidence within the network as a whole for nonrandom patterns of association between animals or groups of animals. Our aim is to use a social network to identify inhomogeneities in the social structure that may be important influences on (or consequences of) the genetic makeup, habitat use, information, and disease pathways of a population (Lusseau et al. 2006), and to relate any such structure to variations in known attributes of the individual animals.

The types of questions we are asking in this chapter are:

1. Does the network as a whole show evidence of segregation by discrete categories such as sex?
2. Is there any tendency for phenotypically similar animals to be bunched together, in a social sense? Do the larger animals in a population hang around predominantly with other large animals, for example?
3. Does the network structure itself suggest that there are groups of animals that are much better connected to each other than they are to the rest of the population? If so, is the whole population structured this way, and how do we try and unearth what drives this segregation?

Such questions have been asked by social scientists for a long time, and there are many methods that have been developed that provide answers or partial answers to these types of question. There is a large literature on this topic, which will, in all probability, find natural extension in the analysis of many animal social networks. It is beyond the scope of this book to give a detailed description of methods covered so fully elsewhere. To get the interested reader started, we suggest you take a look at one of the many excellent books on human social network analysis, such as Scott (2000), Wasserman and Faust (1994), or Carrington, Scott, and Wasserman (2005). A search for key terms such as "roles" and "positions," "block models" and "structural equivalence" should help you on your way. Other methods more familiar to the biologist, such as hierarchical clustering (see Kaufman and Rousseeuw [1990] for example) may also be turned to such questions; we will come back to this a little later.

Our approach in this chapter will be to look to some of the recent advances in network analysis, often developed by those with a primary interest in the physical sciences, which might be brought to bear when tackling such structural questions. It remains to be seen what the relative merits of these various approaches will turn out to be. A subsidiary aim of this chapter is to introduce the reader to a few of the ongoing developments in network analysis that are happening in a wide range of disciplines, and to encourage an open-minded view as to where to look for help. Again, our intention is to be exploratory and suggestive, rather than exhaustive, in our treatment of these issues.

We will try to show you how to address these questions using two closely related, but distinct, approaches. Questions 1 and 2 may be addressed by making an informed guess as to what phenotypic, behavioral, or ecological factors are likely to be shaping the social relations in the system in question, then looking to see whether, statistically speaking, the network exhibits the hypothesized "assortative mixing" (section 6.1). Many of the factors likely to be considered have already been mentioned in previous chapters, and include phenotype (such as species, size, sex, and color), geographical range, relatedness, familiarity, and dominance hierarchies, to name but a few (see Krause and Ruxton 2002). The notion of assortative mixing (or homophily as it is sometimes called in the social sciences), in which individuals are more likely to interact with others of a similar type, is well documented in human social networks. Assortative mixing has been described in humans as a function of race, ethnicity, age, religion, education, occupation, and gender (McPherson, Smith-Lovin, and Cook 2001).

The second method (or rather range of methods, as we shall see) avoids the need to formulate a hypothesis at the outset about what mechanisms, if any, are causing the presence of substructure in the network. Instead the network itself is used as the primary source of information in a search for sets of nodes that are more densely connected among themselves than they are to the other nodes in the network. In network parlance, such entities are referred to as "blocks," "cliques," or "communities," depending on how they are defined. We will concentrate on methods to find "communities" (sections 6.2 and 6.3). We are aware, of course, that ecologists use this word in a rather different context, which we hope will not cause them undue displeasure. Should our search of the network reveal that there are indeed communities in a network sense, and we are satisfied that they are statistically robust entities, then the real work begins in trying to determine, *post hoc* this time, what biological, ecological, geographic, or phenotypic features may have caused this community structure.

The chapter ends with two short sections. The first (section 6.4) gives a brief review of some other methods that achieve the same thing as community detection, or something similar. The second (section 6.5) gives a summary of some of the uses of these techniques so far in the study of animal social structure.

6.1 MEASURES OF ASSOCIATION PATTERNS IN SOCIAL NETWORKS

A common approach used to quantify mixing patterns in social networks is to compute some form of correlation coefficient. An example, which is appropriate in the case when individuals in the network can be split into a number (m) of discrete categories (for example sex and species) is Newman's assortivity coefficient r (Newman 2003b). To find r we first construct the $m \times m$ "mixing matrix" $\mathbf{e}$. Unlike the association matrix introduced in chapter 2, each row and column of a mixing matrix represents a different category. The elements of $\mathbf{e}$ tell us the fraction of edges between different categories. For example, suppose that we wish to look for sex assortedness in an imaginary animal social network with undirected edges. Then we have $m = 2$ categories (male and female). We search through all of the edges in our network and find that 50 percent of them are between two nodes each representing a female, 30 percent between males, and 20 percent between a male and a female. In this case our "mixing matrix" is

$$\mathbf{e} = \begin{pmatrix} 0.5 & 0.1 \\ 0.1 & 0.3 \end{pmatrix}.$$

Before we go on, we should note a couple of things about this matrix. The first is that the 20 percent of edges between males and females have been split equally between the upper right and lower left entries in $\mathbf{e}$; this is just a technical convenience, used to ensure that the sum of all the entries in the matrix, denoted $\|\mathbf{e}\|$, is one. There is no implication that half these edges go from males to females and the other half from females to males. When there are more than two categories, care must be taken to ensure that half of the edges between nodes in category s and t are placed in the element e_{st} and the other half in e_{ts}. Also, it is worth pointing out that, since the matrix $\mathbf{e}$ represents the fraction of edges between nodes within and between different categories, it is perfectly permissible to consider networks with weighted, as well as unweighted, edges, and networks with more than one component. The case of directed networks is a little more complicated, and is not covered here. There is no reason in principle, though, why the methods presented in this chapter cannot be used to analyze directed networks.

A simple method to measure the assortivity from the mixing matrix may already have sprung to mind. Surely we could simply add up all the terms along the leading diagonal (the one that runs from top left to bottom right)?

In matrix-speak, a language in which our reader is not assumed to be fluent, the sum of these terms is called the trace of a matrix, and is written as Tr **e**. In a mixing matrix it represents the total fraction of edges between nodes in the same category; for the **e**-matrix of our fictional animal society it is 0.8, so 80 percent of all edges are "within category," suggesting a highly assorted network. However, the trace on its own is an unreliable measure of mixing because it doesn't take into account the possibility that the number of nodes in each category might be widely different. Suppose for example that our fictitious animal population contains 90 females and 10 males; the fact that 20 percent of the edges are between a male and a female might be more important than a simple calculation of Tr **e** would suggest.

A better estimate of assortative mixing between node categories is the extent to which the number of intra-category edges is more than would be expected if the edges were arranged randomly with respect to category. There are various measures (Newman 2003a) that estimate this "excess assortment." One of the more reliable seems to be Newman's assortivity coefficient r (Newman 2003b), which uses the sum of the rows and columns in a mixing matrix **e**. The sum of the elements in the first *row* of the **e** matrix for our fictitious network gives the fraction of edges (0.6) that start at females and end at nodes of either category. The sum of the elements in the first *column* gives the fraction of edges that end at females, having started at a node of either category (0.6 again, as we have an undirected network). If we were to construct a random network with the same number of edges, but with the edges placed with no regard to the *category* of the nodes they connect, we would still expect a fraction 0.6 to go from females, and 0.6 to females, so we should expect the fraction of edges from females to females to be $0.6^2 = 0.36$. Newman's assortivity coefficient measures the excess assortment by subtracting the expected value of the fraction of intra-category edges (0.36 for female–female edges) from the true fraction (0.5), and sums these excesses over all categories. The answer is normalized (forced to have a value less than or equal to 1) by dividing the sum of the excesses by the maximum excess for a perfectly assorted network. Using a_s to denote the sum of row s of **e**, the assortivity coefficient may be written in at least two forms, including

$$r = \frac{\sum_{s=1}^{m}(e_{ss} - a_s^2)}{1 - \sum_{s=1}^{m} a_s^2} = \frac{Tr\ \mathbf{e} - \|\mathbf{e}^2\|}{1 - \|\mathbf{e}^2\|}, \tag{6.1}$$

where the second version of the formula, used by Newman, illustrates how r is different from a simple trace of **e**. If there is no assortative mixing, there are no excess edges, and we see from equation (6.1) that $r = 0$. If there is perfect assortment, only the diagonal elements of **e** are nonzero, so $r = 1$. For our simple fictitious population, $r = 0.583$.

The next issue, of course, is to test whether a measured value of r is statistically different from zero. For this we need to look at the data that gave rise to the measure, which is of course the network itself. For our fictitious case we do not have a particular network in mind, so we will leave that result ($r = 0.583$) untested. But how should we test for statistical significance of assortivity in a real network? Well, if there are doubts as to whether the network we have is truly representative of the social relations under investigation, a full-blown randomization of the experimental data may be needed, of the type we advocated in section 5.3. If we are confident that the network we are analyzing is truly representative of the interactions or associations involved, something more straightforward may be used. For example, we could randomize node labels to generate the expected frequency distribution for our test statistic, as we did in section 5.2. However, in the present case, we can avoid randomization altogether. This is essentially because our test statistic (r) is derived from a mixing matrix $\mathbf{e}$, not from the full network. We can treat each of the E edges in our network as independent contributors to the elements of $\mathbf{e}$, so the usual problem of interdependence of the data disappears. Newman (2003b) recommends the use of a jackknife procedure (Efron 1982) to compute an approximate variance in the value of r:

$$\sigma_r^2 \approx \sum_{i=1}^{E} (r - r_i)^2,$$

where r_i, is the value of the assortivity coefficient r calculated for the network with the single edge i removed. This is still a computationally intensive procedure to perform, but less so than some of the randomization tests of Chapter 5. (As an aside, the reader could be forgiven for wondering why the jackknife procedure was not invoked in chapter 5, when we were looking to test node-based measures. The simple answer is that to find, for example, whether a mean value of the clustering coefficient C of a network is significantly large, each edge *cannot* be treated as an independent contributor to the test statistic, so we cannot estimate the variance in C by jackknifing one edge at a time. A similar argument holds for both the mean path length and the betweenness, though there may be an argument that a mean degree k could be tested this way.)

Newman (2003b) uses the jackknife method to test for significant assortative mixing in example human social networks, and also discusses the merits of one or two other measures, such as that proposed by Gupta, Anderson, and May (1989). In one of the few uses we have seen of this approach in animal social networks, Wolf et al. (2007) used Newman's coefficient r to analyze homophily in a social network of Galápagos sea lions (*Zalophus wollebaeki*) using $m = 6$ categories based on the age (pup, immature, and adult classes) and sex of the individuals. In the overall network, individuals were positively

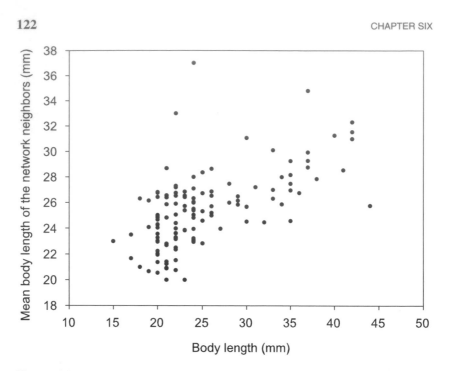

Figure 6.1. The mean body length of the network neighbors of an individual guppy plotted against the body length of that individual. From Croft et al. (2005).

assorted by age class, but assortment by sex was negligible. Considering assortment by sex for each age class separately, immatures were significantly sex-assorted, while adults and pups were not (Newman's $r \pm CI_{95\%}$: immatures: 0.13 ± 0.093, $P < 0.05$; adults: -0.08 ± 0.13, $P > 0.05$; pups: 0.02 ± 0.07, $P > 0.05$). The authors set these results in the context of further investigation into the substructure of the sea lion network (see section 6.5).

Individuals in a network may also assort by more continuous (scalar) phenotypic variables such as age and size. When applied to continuous categories, Newman (2003b) points out that the coefficient r becomes Pearson's correlation coefficient r_p (Sokal and Rohlf 1994). There are other techniques that can be used in the case of continuous variables, however. For example, in an investigation of guppies (*P. reticulata*) Croft et al. (2005) investigated the degree of assortative mixing by body length in their networks by plotting the body length of an individual against the mean body length of its network neighbors. The results for one of the four populations analyzed are shown in figure 6.1. Croft et al. (2005) measured the correlation using the Spearman rank correlation coefficient r_s, and found strong evidence for assortative mixing in the network as a function of body length ($r_s = 0.59$, $n = 130$, $P < 0.001$). (In retrospect, we think that these authors (us!) should have been slightly more careful with this

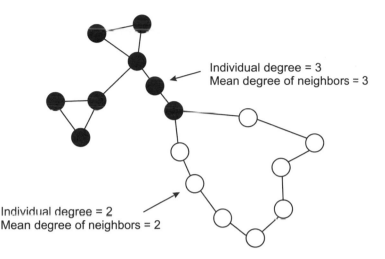

Individual degree = 3
Mean degree of neighbors = 3

Individual degree = 2
Mean degree of neighbors = 2

Figure 6.2. A simple network exhibiting positive degree correlation. Each filled-circle node has degree 3, and each open circle node degree 2. The vast majority of edges in this simple case link like-degree nodes. From Croft et al. (2005).

analysis, as the data points are not independent. In the spirit of Dugatkin and Wilson (2000), they should perhaps have compared the correlation coefficient (the slope of the line) with those generated from a series of randomizations of their data. You can be assured that the authors are duly chastened. Happily, none of their findings are affected.)

Degree Correlation

One form of scalar assortative mixing that has received a fair bit of attention in the networks literature is assortative mixing by degree, the number of network neighbors an individual has. The question here is whether, on average, individuals with high degree tend to be connected to others with high degree. Figure 6.2 shows a simple example of a network with positive degree correlation. Assortative (or indeed disassortative) mixing by degree is again investigated by searching for correlation. Newman (2003b) used the Pearson correlation coefficient as a convenient single measure of the correlation between the degree of an individual and the average degree of its neighbors.

So what might be the significance of a positive, negative, or zero degree correlation? There appears to be no general answer to this yet, but it is interesting to note that social networks appear to be rather different from other types of network in this regard. For example, metabolic networks, food webs, and neural networks usually exhibit a negative degree correlation (Newman 2003a), while many social networks have a positive degree correlation (i.e., well-connected people know other well-connected people). This has been documented many times in human social networks, such as coauthorships of scientific papers and e-mail address books (Newman 2003a), but it also appears that social networks in animals may show similar features. For example, Lusseau et al. (2006) found $r_p = 0.170$ for a social network of bottlenose dolphins (*T. truncatus*). Croft et al. (2005) reported degree correlations of between 0.28 and 0.7 (all of which were significant) for 5 populations of small freshwater fish.

The mechanisms underlying the formation of these patterns, and the apparent differences between social and other networks, are not yet fully understood and provide an interesting field for further investigations. It is certainly a potentially interesting result to find that well-connected animals tend to be connected to each other, and indeed that animals with few connections tend to be found together. One confounding factor that needs to be considered, though, is the effect of sampling methods on the degree correlation. For example, if the network has been constructed via the "gambit of the group" (see chapters 2 and 5), with an edge placed between all animals seen in each group, any variance in the group sizes will tend to produce a bias toward a positive degree correlation. One extremely large group will give rise to a large number of animals that are connected to at least the number of other animals in that group. Thus again, we are led to consider the importance of filtering our networks (chapters 3 and 5) before we draw too many conclusions from their analysis.

6.2 COMMUNITY STRUCTURE IN ANIMAL SOCIAL NETWORKS

Our aim in this section is to discuss methods that might be applied to a network to identify a particular type of heterogeneity in social interactions or associations by searching for "communities," by which we mean collections of animals that are associated more with each other than they are with the rest of the population. Any such community would represent an intermediate level of social organization, above the dyad or group and below the population. By relating community membership to known phenotypes, characteristics, or ecological constraints, for example, we may better be able to understand the interplay between biological, geographical, and other factors and the patterns of social organization in a population, and so start to unravel elements of the structure that would be difficult to see at the level of the dyad, the group, or indeed at the population level. If a key property is found to vary significantly between communities, then it is likely to be misleading simply to present a mean or median value of that property over the whole network population. Community detection has been used in recent years in other biological contexts to search for, among other things, functional modules in metabolic networks (Guimerà and Amaral 2005), and a few examples have now been seen in the study of animal behavior, as we shall see.

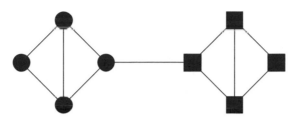

Figure 6.3. A simple network consisting of two communities (those with circular nodes and those with squares) with just a single edge between them.

Previous work on social structure in wild populations has often considered large animals such as primates, ungulates, and cetaceans (Chepko-Sade, Reitz, and Sade 1989; Whitehead and Dufault 1999; Cross et al. 2004; Cross, Lloyd-Smith, and Getz 2005). Such studies have traditionally investigated community structure (though these authors have not called it that—the use of "community" in this context is another piece of terminology that has arisen in the wider networks literature) using clustering algorithms (see Kaufman and Rousseeuw [1990] for an example) to clump together animals that are most alike according to some chosen measures of similarity. In more recent examples of cluster analysis, Lusseau et al. (2003), Vonhof, Whitehead, and Fenton (2004), and Wittemyer, Douglas-Hamilton, and Getz (2005) have used relational data in the form of association indices to measure the closeness of individuals, and have detected intermediate levels of social structure of the type of interest here. Their work is much closer in spirit to the network-based methods we are presenting in this chapter. The starting point for these analyses is very similar to ours, as a network is simply a visualization of an association matrix.

So how do we go about searching for communities in social (or any other) networks? We are looking for subsets of nodes that are more densely connected among themselves than they are with the rest of the network. As an illustration, consider the simple network depicted in figure 6.3. There are eleven edges; of these, five connect circles with circles, five connect squares with squares, and only one edge connects a circle to a square. It is easy to see that the circles and the squares make a sensible "partition" (as we will call a particular allocation of nodes to communities) of the network into two communities in this case. The real issue is to create a computational scheme that will spot this as well as we can. For more complex networks, even the human eye can fail us. Using visualization tools such as those available in NETDRAW (see chapter 3), it is very easy to pick out components—groups of nodes completely separated from each other; each one of these is a perfect community in the sense used here. Spring embedding and other layout schemes may well identify tightly knit communities with few edges between them. Nonetheless, there are many communities, especially those with more than just a few edges between them, that cannot be identified by eye, so we need help from a computer algorithm. (Even for relatively straightforward "partitioning" of a network into communities, we would still like to have some idea of how "strong" or well defined they are, which is again difficult without some computational help.)

The problem of finding a set of communities automatically from a network is decidedly nontrivial, though quite a few methods have been proposed. In each case, a systematic search is performed on the network to find the likeliest partition of the members of the population into communities of highly connected individuals. Most schemes are either "divisive" (Girvan and Newman 2002; Radicchi et al. 2004) or "agglomerative" (Newman 2004). Divisive algorithms remove edges from the network that are, based on some prescribed network metric, good candidates for being inter-community edges. This process eventually causes the network to be split into a hierarchy of communities. Agglomerative algorithms, in contrast, start with no edges in the network, and repeatedly add edges that are most likely intra-community edges. Clustering methods, including those used recently to find social structure in bats (Vonhof, Whitehead, and Fenton 2004) and elephants (Wittemyer, Douglas-Hamilton, and Getz 2005), are an example of an agglomerative method for determining community structure.

The Girvan and Newman (2002), henceforth GN, algorithm is a divisive scheme that has become one of the more popular methods for detecting communities in networks, perhaps in part because it is readily available for use; it is, for example, an analysis tool within the NETDRAW program. The GN algorithm uses edge betweenness (chapter 4) as a measure to determine which edges are most likely to join communities. You will recall that the betweenness of an edge is the number of shortest paths between nodes that make use of that edge. It provides an intuitive method of determining which edges lay between communities. In the simple network in figure 6.3, the edge connecting a circle to a square has a much higher betweenness than any other, so it would be the first to be removed by the GN algorithm, splitting the network into two communities at the first step. All that is needed to complete the GN approach is a measure of the "strength" of the partition of nodes into a set of communities, which is used to stop the process of edge removal before they are all gone and we are left with as many communities as there are individuals.

Newman and Girvan (2004) introduced a stopping parameter that has been used widely since. They called it the modularity, Q. At any partition into g communities, Q is defined (on the original network, with all its edges present) as

$$Q = \sum_{t=1}^{g}(e_{tt} - a_t^2),$$
(6.2)

where e_{tt} is the fraction of edges within community t, a_t is the fraction of edges with one or both ends in community t, and the summation runs over all g

communities. Q measures the extent to which edges between individuals are intra- rather than inter-community. The "best" partition of the nodes into communities is that which gives the highest value of Q.

The keen-eyed reader will have noticed the striking similarity between Q and Newman's assortivity coefficient r (equation 6.1). Apart from the fact that Q includes no normalization factor, the two measures are very similar indeed. The real difference is that when using r, we are partitioning the nodes into predetermined categories; when using Q, the nodes are partitioned into a (possible) set of communities based only on the topology of the network edges. It will perhaps be helpful at this point to highlight that the search for communities is not therefore the same as a search for homophily. Finding a significant assortivity coefficient doesn't indicate whether a given category of animals is in one small region in the network or dispersed as topological cliques. To answer that, we need a community analysis, or something similar. Equally, a low value of r simply means there is no homophily with respect to the categories chosen. In a search for communities, we are allowing the network structure itself to determine structural heterogeneity, rather than choosing categories at the outset by which we expect the population to segregate. Thus we may find assortment by categories that never occurred to us at the outset. A significant community structure may, or may not, be consistent with significant assortivity by any category in the network as a whole.

As with the construction of r (section 6.1), care must be taken only to count each edge once in the evaluation of Q, and this can again be achieved by assigning half of each edge between communities s and t to e_{st} and half to e_{ts}. Q takes a value of 0 in the case of a random assignment of community structure, and can in principle be negative. We are much more interested in positive values; in practice, Q is rarely larger than 0.7 (Clauset, Newman, and Moore 2004), unless communities are virtually separate components, in which case finding them is very easy. For the toy network in figure 6.3, $Q = 9/22 \approx 0.41$.

The GN algorithm has great appeal due to its intuitive nature and ready availability. It has been used by Lusseau and Newman (2004) and Lusseau et al. (2006) to find community structure in dolphin networks, as we will discuss further in section 6.5. However, as we will show in section 6.3, there are cases where it cannot detect communities that are there and are statistically meaningful. Such a scenario is most likely in fission–fusion systems, where the rapid turnover of group membership gives rise to more than just a few inter-community edges. Then there can be partitions into communities with higher values of Q than the GN algorithm can achieve, which may, if statistically corroborated, reveal more detail of the social structure than would otherwise be found.

The methodological problem that leads to this scenario is that both agglomerative and divisive methods (including the GN algorithm) suffer from being inherently "one-way." What we mean by this is that once an edge has been removed (for divisive algorithms), it cannot be put back again. This has the effect of removing potentially better partitions of the graph from consideration early in the execution of the algorithm. So how might we solve this? Well, one answer would be to search for the node partition with the highest Q simply by trying them all. If we manage to do so, all this "one-way" business will be irrelevant; if all possible parts of "partition space" have been visited, the highest Q we find will, definitively, be the highest possible Q.

Of course, it is easy to see that such a process isn't simple at all, at least in the execution. Remember that at the outset, we don't know how many communities there are, or how big they are, so we must explore all possible combinations of different nodes being allocated to different numbers of different-sized communities. For each allocation we use equation (6.2) to evaluate Q, and look for the allocation with the largest Q. Even for the very simple network in figure 6.3 there are a lot of possible partitions to try (box 6.1). What we need instead is a method of homing in on the best partition without trying every single one. There is no foolproof scheme to perform such "combinatorial optimization" (and nor is there a very snappy name for it), but one which often does well is "simulated annealing."

Box 6.1

Counting the Number of Partitions

To find the total number of distinct possible allocations (or "partitions") of nodes to communities, we are really asking the question, "How many ways can you split up n distinct objects into groups?" Let's try to answer this question in two stages for the simple $n = 8$ network in figure 6.3.

First, we want to calculate how many different combinations of group sizes there are for n objects. We can get help here from number theorists, some of whom care deeply about the number of ways you can add whole positive numbers to arrive at a given total n. They call this the "partition number" of n. (This is exactly what we need, but perhaps confusingly, it is a slightly different use of the word "partition" from the one in the main text, where a partition is any split of the network into communities.) The partition number of 8 is 22—this is not too hard to see. You can have one group of 8, one of 7 and another of 1, 2 groups of 4, and so on. Exhaust all the possibilities and you get to 22.

Now for stage two. We need to know, for example, how many ways there are to get one group of 7 and one of 1. Well, there are 8—each of the 8 nodes could be the one on its own. We need to do the same for all 22 splits of 8 objects into groups (taking care not to count the same solutions more than once—this is the tricky bit). Some splits generate many more possibilities than others. For example, there is only one way to get a group of 8, but 840 ways to get a 3, two 2s, and a 1.

So the total number of community partitions is found by adding together all the distinct solutions from the 22 possible sets of group sizes. We found 4,140 distinct possibilities. This may not sound too bad, but things rapidly get worse as the network gets bigger. For example, the "partition number" for $n = 100$ is 190,569,262, and there will be many more ways in which each of these can be configured. So we hope it is clear that an exhaustive search of all possible partitions soon becomes a prohibitively intensive computational exercise.

Simulated annealing is a well-established optimization strategy often used in combinatorial problems (Kirkpatrick, Gelatt, and Vecchi 1983). Recently it has been used to find community structure in metabolic networks (Guimerà and Amaral 2005). Unlike agglomerative and divisive schemes, simulated annealing allows us temporarily to consider poorer quality communities (those with relatively low values of Q) in order to provide a means of escaping a region in "community space" of locally, but not globally, good communities (see box 6.2). The solution space of the problem can, initially, be explored quite freely. The accessibility of the lowest quality parts of the space is gradually decreased, and the partition converges (we hope) on the global optimum of the problem. The additional sensitivity that this brings to the algorithm is illustrated by the fact that we can, in general, obtain a value of Q for a given network that is at least as large as for any other community-finding algorithm of which we are aware. The cost to be paid for this accuracy benefit is that simulated annealing is a relatively slow means of finding network communities, though this will only be a problem if there are more than a few hundred animals in the data set.

Since our aim is to be able to look for communities in highly dynamical social systems, it is essential that we augment our search with statistical tests of whether the partition we find is meaningful (there is no reason to believe that all social systems should contain intermediate levels of organization), and others to determine what factors may be driving the partitioning of the population into communities. We will illustrate these issues via a case study.

6.3 CASE STUDY: COMMUNITY STRUCTURE IN THE TRINIDADIAN GUPPY

We will use as a case study the social network of a wild population of 197 Trinidadian guppies (*Poecilia reticulata*) caught, marked, and recaptured

Box 6.2

Simulated Annealing

Instead of maximizing our parameter Q, let us minimize $E = -Q$. At the heart of simulated annealing is a series of trial moves in the space of all possible allocations of nodes to communities, which take the community allocation from its current state (characterized by E_{before}) to a putative new state (E_{after}). Let $\Delta E = E_{after} - E_{before}$. A trial move is accepted unconditionally if ΔE is negative. If ΔE is positive, the move is accepted conditionally, according to a probability distribution:

$$\text{p(accept move)} = \exp(-\Delta E/T).$$

In other words, moves that worsen E may be accepted, but this is much more likely if the degradation is small. The name "simulated annealing" derives from an analogy with annealing a piece of metal by a careful reduction in its temperature to produce the lowest energy crystalline metallic state. In annealing a metal, E would represent the energy of the system and T the temperature. In this application T is a control parameter used to govern how much of the solution space is available to be explored (i.e., how far "uphill" in E a move can take us). It should be set such that almost all moves are accepted at the start of the annealing process. As the search proceeds, T is systematically reduced (the system is "cooled") to allow the allocation of nodes to communities to settle on what we trust (though cannot prove) is the partition with the lowest possible E, and thus the highest possible modularity Q.

The art of simulated annealing is in choosing the "cooling" plan and the trial moves that adequately explore the phase space of the problem in hand. For the guppy example in section 6.3 we consider three types of trial moves, chosen empirically to suit network problems:

1. A node is chosen at random, and assigned to a randomly chosen community.
2. Two nodes are chosen randomly, and swap communities.
3. A node is chosen at random and assigned to the community of one of its immediate network neighbors.

Each type of move is chosen with equal probability. The trial moves allow us to move in Q-space to other points that are close by.

In order to choose a sensible initial value of the control parameter T, 1,000 random moves are performed, yielding a mean ΔE per move; T is then chosen so that 95 percent of these moves would be accepted. An empirically chosen $5ng$ moves (n is the number of nodes, g the number of communities) are performed at each value of the control parameter, before reducing its value by 5 percent. The calculation is taken as converged once there is a negligible change in $E(-Q)$ over the preceding five values of T.

in two adjacent pools (henceforth the "upper pool" and "lower pool") on a stretch of the Arima River, Trinidad (see figure 6.4). The guppy is a very-well-studied model system with frequent fission and fusion of shoals, whose social networks we have used before in this book. The analysis in this section was conducted by David Mawdsley.

Our starting point for the search for communities in guppy societies is the social network depicted in figure 6.5, which has been filtered (see chapters 3 and 5) to include only those pairs that were observed twice or more together. Note that the networks in figures 6.5 and 6.6 have been laid out using a spring-embedding algorithm (chapter 3), and not using any spatial information on where the fish were caught. For further details, including the protocol used to collect the data, see Croft et al. (2005). Our aim here is to find evidence of statistically meaningful community structure within this network of connections. The example we have chosen rather nicely illustrates one obvious, easily detected partition (caused by the geography of the two pools), but also a far-from-obvious partition into two communities within each pool, which we find to be statistically and biologically meaningful, and consistent with phenotypic assortment of the population at the level of the community, intermediate between the shoal and the population.

Identifying Statistically Significant Network Communities

Applying the simulated annealing scheme outlined in box 6.2 to the guppy network in figure 6.5 yields the community structure shown in figure 6.6, in which the communities are coded by node shape. There are five communities in the network. The value of Q for this partition is 0.556, which is relatively high; networks with a strong community structure tend to have Q in the range 0.3–0.7 (Newman and Girvan 2004). Although a necessary criterion of community structure, a high value of Q is not always indicative of any *meaningful* community structure. Guimerà, Sales-Pardo, and Amaral (2004) point out that many random networks, with no obvious "communities" in the sense discussed here, can have rather high values of Q, so it is important that our result be subjected to a statistical test of significance. Such tests are not routinely presented as part of explicitly network-based community detection; however, without them it is hard to draw any firm conclusions about community membership unless the partition is very clear cut. In strongly dynamical systems,

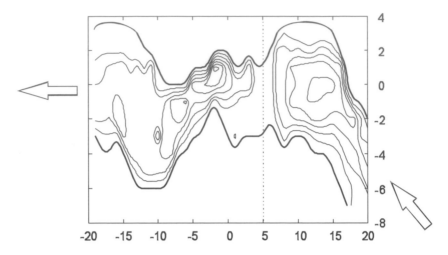

Figure 6.4. A rough map of a 40-meter stretch of the Arima River, Trinidad. Arrows indicate the direction of water flow and contours the water depth. There are two pools, separated by a shallow riffle in the region of the dotted line. Note that the vertical and horizontal scales are not the same.

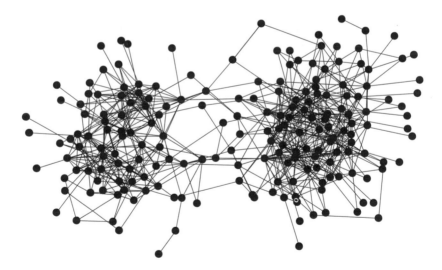

Figure 6.5. The social network of Trinidadian guppies taken from two adjacent and interconnected pools in the Arima River. Each filled-in circle represents an individual fish. Individuals connected by an edge were found at least twice in the same shoal.

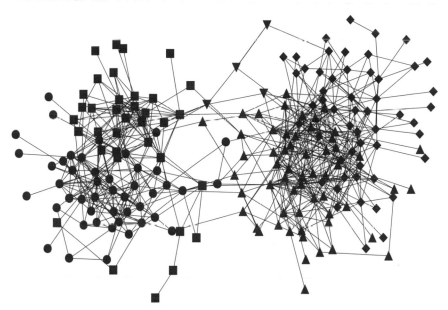

Figure 6.6. The same network as figure 6.5, using the same layout. This time the node shape indicates which of the five communities the fish was placed in (●, ■, ▲, ▼, and ♦) by a simulated annealing program.

where there may be many social relations (edges) between some putative communities, the partition must be tested.

The approach we have adopted to test for the significance of Q is a Monte Carlo test (see chapter 5 or Manly [1997]). A randomized version of the data is produced, conserving the daily recapture frequencies and groups sizes (Ward et al. 2002). From the randomized data a randomized network is produced and filtered, and the simulated annealing algorithm used to find the communities with the largest Q-value for the randomized network. This process is repeated 1,000 times to yield a P-value by comparing the expected value of Q from the randomized networks with our measured value. Comparing the Q-values for 1,000 random realizations of the experimental data in this case yields $P <$ 0.01 (only 8 of the 1,000 randomizations gave a higher Q than the observed value $Q = 0.556$), so we conclude that the community structure in figure 6.6 is statistically significant.

Factors Influencing Community Structure in Guppies

The question to ask now is whether we can attach any biological significance to the membership of the communities we have found. The analysis here is necessarily rather specific to the study system. In the following we will concentrate on the four large communities that provide us with sufficient data for more detailed analysis (figure 6.6), and refer to these communities by node shape as ●, ■, ▲, and ♦. The fifth community (▼) contains just four individuals.

The obvious separation of the network is into two communities (▲ and ♦) associated with fish found predominantly in the lower pool in the study site, and two (● and ■) associated with fish found mostly in the upper pool. To confirm this simple interpretation, we looked at the position along the river (see figure 6.4) of each recaptured shoal and constructed as a test variable the median position along the river at which each fish appeared during the fifteen days of the experiment. Most fish are found to be wide ranging, but the vast majority of recaptures are clearly in one pool or the other. Combining fish in communities ▲ and ♦ (n = 82, median x_{50} = −13.6 m, inter-quartile range (IQR) 5.4 m) and those in ● with ■ (n = 111, x_{50} = +11.2 m, IQR 1.7 m) yields a significant difference in position (Mann-Whitney U (MWU) test; z = −11.4, P < 0.001), confirming that inter-pool social ties are rare. We note in passing that the GN community-finding algorithm of Newman and Girvan (2004) makes little progress with this network beyond the split between pools, and yields a lower Q-value (0.467) than we found via simulated annealing (0.556). This is simply a manifestation of the "one-way" nature of the GN algorithm, as we discussed above.

The most interesting comparisons are between the communities associated with each pool, since it is far less obvious from a casual inspection of the network that each pool contains two communities. Owing to the large movements of the fish throughout a pool over the course of the experiment, their median capture position is probably not a reliable test variable. We looked for differences between the distribution of body length of fish in the communities, their median water depth of capture (which is more reliable than position, as many regions have similar depths) and the proportion of females they contain.

Within each pool the two communities show significant differences in both median body length (figure 6.7) and median depth of water at capture (not shown). There is less evidence for sexual segregation between same-pool communities, with the distribution of males and females between communities not differing from that expected from an even distribution in both the upper and lower pools (upper pool—community ●: females = 37, males = 21; community ■: females = 27, males = 26; Chi Sq test: χ^2 = 3.79, d.f. = 1, P = 0.052. lower pool—community ▲: females = 33, males = 13; community ♦: females = 27, males = 7; Chi Sq test: χ^2 = 1.38, d.f. = 1, P = 0.240). However, it should be noted that in the upper pool the result was only marginally non-significant, suggesting that sex may have some role to play in community structure.

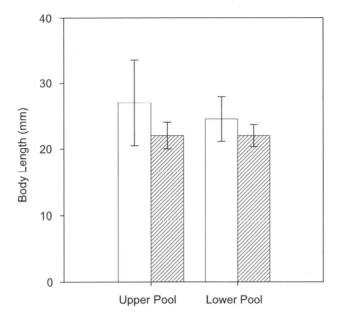

Figure 6.7. Median ($\pm$ the inter-quartile range) body length of guppies within each community of each pool. Upper pool—●: unshaded, n = 58, ■: shaded, n = 53. Mann-Whitney U test: z = −4.52, P < 0.001. Lower pool—▲: shaded, n = 46, ◆: unshaded, n = 36. MWU test: z = −2.47, P = 0.014.

Identifying the Roles Individuals Play in Interconnecting Communities

If it is the case that organization of the population by phenotype occurs at a community level and not at the group level in the Trinidadian guppy, then it becomes an interesting issue to identify those animals at the boundaries between communities, as these hold the network together and may be expected to play a particularly important role in the transmission or inhibition of information through the population. A few of the recent methods of detecting communities (such as that of Reichardt and Bornholdt [2004]) deal with this issue by searching for "fuzzy" communities, in which the allocation of some individuals to a particular community is equivocal. We adopt a simpler approach, which appeals to a heuristic notion of community. For each node in the network we calculate the fraction f of its edges that connect to other members of its own community. We arbitrarily denote all nodes with $f > 0.7$ as core community members. Of 197 fish, all but 39 are core community members (and 75 have exclusively intra-community edges, with $f = 1$). The 39 peripheral fish are, in a community sense, more itinerant members of the population. In our guppy system we do not have the test power to determine unequivocally whether sex or water depth sets the core and peripheral animals apart, though it appears that body length may be an important factor. The only significant differences are in the upper pool, where the body length of the 20 peripheral community members of ● and ■ combined ($x_{50} = 22.5$ mm, IQR $= 4$ mm) is significantly different from (and lies between) that of the 48 core members of ● ($x_{50} = 30$ mm, IQR $= 12$ mm; MWU test: $z = -2.84$, P $= 0.004$) and the 43 core members of ■ ($x_{50} = 22$ mm, IQR $= 4$ mm; $z = -2.18$, P $= 0.03$).

So What Have We Learned?

This case study identifies two large communities of similar size in each of two pools that are not merely a product of the data collection method. There is strong evidence for communities to be assorted by body length. Given that guppies show sexual size dimorphism with females growing larger than males, we should expect sex and body length to covary to a certain degree, though the results here do not show strong sexual assortment. Only experimental work that assesses community structure of size-sorted fish when both sexes are present can answer the question of whether sex and size are independent factors. We also find that the physical environment plays an important role; natural barriers limit the exchange of fish between lower and upper pools, and within pools there is also an influence of water depth on community structure. Larger fish tend to be found in deeper water, and this is thought to be driven by predation risk (Croft, Botham, and Krause 2004b). We have identified individuals at the boundary between two communities, and shown some (suggestive, but not conclusive) evidence that these animals have a body length between that of either community.

Phenotypic assortment in guppy populations in terms of body length and sex segregation has been previously demonstrated at the group level (Croft et al. 2003; Croft, Botham, and Krause 2004b). So why should we care about community structure in this system? One possibility presented by this analysis is that community structure (and not group structure) is what matters when looking at phenotypic assortment by factors such as body length and sex. If this were the case, then interactions at the group level might have quite a different significance depending on whether groups belong to the same or to different communities. It may well be that individuals are exchanged freely between groups provided that they belong to the same community, and that community membership largely determines group composition. Only by looking at whole network of interrelationships can we hope to answer such questions.

6.4 RELATED METHODS

Some of the other methods by which the communities may be found in a network have already been mentioned, including the well-known Girvan and

Newman (2002) algorithm. Community detection continues to excite a good deal of interest, and for a while new methods of achieving the job seemed to appear in the physical sciences side of the networks literature every month. It would be foolhardy and misplaced to try to present a thorough review of these methods here. Instead we refer the reader to Newman (2004) for a review up to 2004, and highlight one or two approaches that have emerged since then that seem to be attractive. One feature they all have in common is an attempt to analyze how strongly an individual belongs to a given community.

Reichardt and Bornholdt (2004) developed a method very much centered in the traditions of statistical physics, by re-couching the problem as one of "energy minimization" with an energy function reflecting whether local network structure is, or is not, consistent with membership of a community. The minimization is repeated many times; nodes that are almost always in the same community are considered core; those that turn up in different communities are "fuzzy" community members. Reichardt and Bornholdt also discuss the need for statistical testing of the communities, which they do by comparing with results for Erdös-Rényi random networks.

Newman (2006b; 2006a) has himself produced another method to detect communities that should be mentioned. It is based on repeated splitting of the network into two (bi-partitioning), based on an analysis of the eigenvectors of a "modularity matrix" derived from the association matrix. The method is one of the best we have seen in terms of the combination of accuracy and speed. A by-product of the method is that it yields a measure (coined the "community centrality" by Newman [2006b]) of the strength of community membership for each individual. This and the fuzzy membership of Reichardt and Bornholdt (2004) are just slightly more sophisticated variations on the theme developed in section 6.3 to look at core and peripheral members of intra-pool guppy communities.

All of the methods we have outlined so far are based on the two central tenets that there is a reliable metric of "good communities" (usually the modularity Q) that should be optimized, and that all nodes belong to precisely one community. On the former, Fortunato and Barthélemy (2007) have shown that there are limits to the reliability of the modularity, though these are very unlikely to be a problem for the size of network of interest to us. As far as the latter tenet is concerned, it is clear that we humans consider ourselves to be part of more than one community, associated with work, leisure, and so on. Palla et al. (2005) present an interesting and different approach to community detection that expressly allows "overlapping" communities, with individuals belonging to more than one community. They build their definition of a community around the "k-clique," which is a perfectly connected group of k nodes. They then define a "k-clique community" as the union of all k-cliques that are joined together by sharing $k-1$ nodes. The authors consider networks with weighted edges and, echoing our discussions in chapter 5, they recommend filtering with an edge-weight threshold, and trying a sequence of thresholds and values of k. As we noted in chapter 5, they develop a rule of thumb for the choice of these two numbers (which they acknowledge are otherwise arbitrary) driven by a view that networks close to breaking up into components have the richest structure. Finally, Palla et al. (2005) use randomized networks (though these have little structure) to check whether their communities seem robust. Though it could be argued that its main selling point, the production of over-lapping communities, is probably only going to tell us something we could learn from other methods by looking at the strength of community member-ship, this paper strikes many other resonances with our own work and views on how to go about exploring network structure.

Of course, the developments that we have chosen to highlight in this chap-ter have continued in parallel in other fields. There is a long history of find-ing groups or clusters in data. In the social sciences there are a whole raft of methods, as we have already mentioned, based on block modeling and similar ideas (see e.g. Wasserman and Faust 1994), and these may well bear fruit in the study of animal social networks. A method that has been already been used rather more in biology is that of hierarchical clustering (Kaufman and Rous-seeuw 1990), in which animals are placed together in a hierarchical organiza-tion of the data based on their similarity, as measured by some metric. The hierarchy is presented as a "dendrogram," or tree structure, with individuals at the ends of the smallest branches and the population at the trunk. If the met-ric is relational, such as an association index, then even if a network is never drawn, the resulting hierarchical structure is probing essentially the same thing as a network community analysis as discussed in this chapter. All that needs to be done is to choose (sometimes slightly arbitrarily) where to cut the dendro-gram to reveal the clusters (communities) of closely associated individuals.

6.5 APPLICATIONS TO ANIMAL SOCIAL NETWORKS

As we have just hinted, the idea of looking for "intermediate-level structure" in a population is not new. The idea has been reinvented many times over in various disciplines, and some of the approaches have been used to look for social structure in animal populations. Whitehead and Dufault (1999) give an excellent and thorough review of many studies across a range of taxa that have used association data to search for structure of essentially the type discussed in this chapter. The methods used include hierarchical clustering and principal coordinate analysis. Whitehead and Dufault (1999) assess the merits of dif-ferent schemes and suggest which approach might best be used for a given system, depending on the number of animals involved.

There have been more uses of clustering analysis, based on association indi-ces, since Whitehead and Dufault (1999) published their review. For example,

Lusseau et al. (2003) constructed a dendrogram of the associations among bottlenose dolphins (*Tursiops* spp.) in Doubtful Sound, New Zealand, and found three mixed-sex groups of animals (communities in our parlance) that tended to be found together more often than they were with others. The makeup of these groups was suggestive of a unique social structure in this population, which the authors proposed is due to its geographic isolation from other populations.

Cross et al. (2004) used a cluster analysis of associations between African buffaloes (*Syncerus caffer*) in the Kruger National Park over a number of observation periods, and found that buffalo herds are not as well defined as had previously been assumed. The clustered social structure is used to inform a dynamical social network model of disease dynamics. Cross, Lloyd-Smith, and Getz (2005) discussed the advantages of a new association index (the fission decision index—see box 3.3) over other indices when determining the dendrogram of fission–fusion populations.

Wittemyer, Douglas-Hamilton, and Getz (2005) used cluster analysis based on associations to identify four social tiers in a population of African elephant (*Loxodonta africana*) and investigated the effects of season, study period, and age on the structure and cohesion of these various tiers. Vonhof, Whitehead, and Fenton (2004) identified a novel social structure among bats in a mark-recapture study of Spix's disc-winged bats (*Thyroptera tricolor*). They found very-well-defined, mixed-sex social groups (communities) with very little interconnection, despite the fact that the groups' spatial ranges were often overlapping. The authors suggest that community membership is based on kinship and cooperation in this population.

Rather fewer studies have so far made explicit use of the type of community analysis presented in this chapter. In one of the first, Lusseau and Newman (2004) used the GN algorithm to identify communities and subcommunities within the Doubtful Sound population of bottlenose dolphins, and related the membership to sex and age-related assortivity. They also identified individuals that may act as links between communities and which appear to be crucial to the social cohesion of the population as a whole. A different population of bottlenose dolphins, off the east coast of Scotland, was studied by Lusseau et al. (2006). They found two social units with rather limited connections between them. By probing the social structure at different length scales, the authors were able to conclude that the two communities were built on genuine social affiliation rather than simple spatial separation.

Wolf et al. (2007) conducted an analysis similar to the one in section 6.3, using simulated annealing and randomization tests, on an island population of Galápagos sea lions (*Zalophus wollebaeki*). The authors report significant community structure, which was largely explained by space use. However, they also found significant subcommunities within each community whose membership could not be explained by space use, male territoriality, or assortment by sex or age. The likely candidates to explain the existence of these subcommunities are genuine social affiliation, genetic relatedness, or a combination of both.

6.6 CONCLUDING REMARKS

The message of this chapter is that many animal social networks are likely to contain substructures consisting of well-connected clusters or preferential connections between classes of animal. Happily, there is a range of techniques available to search for substructures of this type. Finding and analyzing them is likely to reveal something interesting about the system in hand that might have been difficult to find by other means. As we noted in the previous chapter, none of the structures we find can be taken at face value; they must all be submitted to thorough statistical testing. Finally, we can look ahead to chapter 7 by noting that any model of an animal social network should take structures such as communities into account. If there is considerable variation in a network property within each community, a single measure of that property over the entire network is likely to be misleading at best.

7

Comparing Networks

We are quite close to the end of our exploration of animal social networks, yet so far we have only considered how to analyze the structure of a single network. It is clear that the whole network approach will be much more appealing if we are able to compare networks, and that is the subject of this last substantive methods chapter.

The comparative approach is a very powerful tool in ecology and evolution (Harvey and Pagel 1991), and comparing observations between contexts, populations, and species can provide insight into the proximate and ultimate causations of behavior (Krebs and Davies 1996). Given the success of the comparative approach, and the potential of networks to bridge the gap between the individual and the population, we should expect considerable rewards if we are able to make statistically robust comparisons between empirical networks. We may want to compare networks within or between populations and species, and investigate whether there are structural features that distinguish different kinds of social relations between populations or species, or between different contexts such as environmental conditions.

The topic of network comparisons comes at the end of our book not because we think it is the least interesting or important, but because, particularly when it comes to comparisons between populations or species, there are very few examples in the animal behavior literature on which to draw. In an early review of the literature on the application of social network theory to primate societies, Sade and Dow (1994) commented on a comparison of social network structure between species and suggested that such an approach could offer a way of making broad generalizations about the social architecture of animal societies. Maryanski (1987) compared various contemporary hypotheses on the structure of African ape social organization from a networks perspective, concluding that chimpanzees (*Pan troglodytes*) and gorillas (*Gorilla gorilla gorilla*) have similar structural arrangements in their patterns of social organization. However, the analysis was based on categorizing three types of ties in different contexts as weak, moderate, and strong. The approach did not take into account the structural complexity of the association patterns in networks. Furthermore, the analysis was largely qualitative and did not test the significance of the observed patterns. In order to compare networks between populations and species, we need to test whether structural features of the networks are significantly different. Other authors have used networks to compare a single species between contexts, or to compare species; we shall refer to these examples where we discuss the methods used.

A second reason why this important topic has been left to last is that from a methodological perspective, there are a few issues that are not yet fully resolved. Consequently our account contains at least as many questions as answers. We will explore some of the current options for making quantitative network comparisons. In doing so we will often be taking a lead from the social sciences literature, where network comparisons have a much longer tradition (Katz and Powell 1953), though it remains to be seen how many of the methods that have been developed there will find direct use in the study of animal social networks. At the moment, it is safe to state that some network comparisons are easier to deal with than others.

It may help at the outset to divide network comparisons into two categories:

1. Comparison of two (or more) empirical networks based on (more or less) the same individuals
2. Comparison between networks constructed for different individuals

As is usually the case with classification schemes, this one isn't perfect, as some of the methods used in category 2 are also applied to category 1, for example. We could equally well have chosen to classify the possible categories in terms of whether or not comparison is facilitated by a model of the network structure. This will be a recurrent theme in this chapter, especially as we try to overcome problems inherent to the comparison of networks containing different numbers of animals and interactions.

The chapter is organized as follows. We begin with the simplest case in category 1 above, in which we compare two networks constructed for exactly the same set of individuals. In this case we can use a matrix correlation test (section 7.1) to see whether, overall, the two networks are consistent with each other. We might be interested in whether the network derived from one set of behavioral observations (grooming for example) is consistent with another (aggressive interactions, let's say). The same tests can also be used to test whether directed interactions tend to be reciprocated, and indeed to compare a single network with a matrix derived from some measure of similarity between individuals (relatedness, for example) that is hypothesized to be of biological relevance to the formation of the network. We may extend this last approach further and use partial matrix correlation tests to control for the influence of an attribute variable (such as dominance) on the relationship between two networks.

There may be very good reasons for wanting to compare networks representing the same interaction among the same individuals, but at different times; such comparisons are referred to as "longitudinal network analysis" in the social sciences. Of course, we can't always engineer our repeated observations so that the same individuals are involved. If we want to perform a longitudinal analysis of network structure from long-term studies of social affiliation to look for long-term trends, life-history effects, and so on, then birth, death, immigration, and emigration will mean that we are no longer comparing networks constructed on the same individuals.

This is where things start to get a little trickier. Standard correlation tests are not appropriate when our networks contain different individuals, or in deed when we have more than two networks to compare. The issue of having a variable number of animals is a real one since, as we outline in section 7.2, the number of nodes n and edges E can each have a profound effect on measures of network structure, and these differences threaten to mask any more-fundamental structural differences. To overcome this problem we will explore the possibility of using simple network measures presented in chapter 4 to characterize (and hence compare) network structures. We will see that one of the more promising ways to try to control for the effects of n and E is to use some sort of model of each network. We have already flirted with this approach earlier in the book. In chapter 4 we introduced the notion that standard, analytically tractable networks such as the Erdös-Rényi random network or a regular network could act as a rudimentary benchmark against which to compare mean network measures such as a path length or clustering coefficient. By chapters 5 and 6 we were advocating the use of Monte Carlo methods to test for statistical significance of observed structures. These methods essentially involved the construction of null models, trying to establish those features (of the data collection protocol as much as anything else in those cases) that must be controlled for before we can search for the meaningful biology. In section 7.3 we give a brief introduction to a family of models used by social scientists to characterize the structure of a single network, and illustrate how they have been adapted as tools for network comparison in a longitudinal analysis and between populations, contexts, and species.

7.1 COMPARING TWO NETWORKS MEASURED ON THE SAME SET OF INDIVIDUALS

Let us begin with the most straightforward case we will consider in this chapter, which is the comparison of two networks constructed for the same set of individuals. We will compare the distribution of values in two association matrices (see chapter 2). Recall that in this book we are using the term "association matrix" to denote any set of pair-wise relations, whether biological or physical in origin, and independent of whether the interactions or associations are weighted or directed.

So let us call our two association matrices **X** and **Y**. Each is a square matrix with as many rows and columns as there are animals, and each represents a

TABLE 7.1.
A pair of fictitious association matrices, **X** (left) and
Y (right), representing different networks derived
from the same three individuals, labeled A–C.
Both are weighted but undirected.

	Individual					Individual		
	A	B	C			A	B	C
A	0	4	5		A	0	1	2
B	4	0	6		B	1	0	3
C	5	6	0		C	2	3	0

different measure of interaction or association. The matrix element X_{ij} gives
the strength of the interaction represented within **X** between animals i and j.
A very simple example is shown in table 7.1.

   In order to compare **X** and **Y** we can use a correlation test to investigate
whether there is a tendency for large (or indeed small) values in **X** to be in
the same positions as they are in **Y** (a positive correlation), or if the reverse is
true—i.e., small values in **X** are associated with large values in **Y** (a negative
correlation). If they are, then the matrices are correlated and we would learn
that there is a (correlative) link at the network level between the behaviors represented in the two networks. However, standard methods such as the Pearson
and Kendall rank correlation tests (Sokal and Rohlf 1994) are not suitable because they assume we have independent data points. As we have already discussed in previous chapters, elements in a matrix are not independent because
the same individual occurs repeatedly, and this creates dependency within
and between rows and columns. For a detailed discussion of the problems
associated with the statistical analysis of matrices, see Douglas and Endler
(1982). To make comparisons between matrices we need to use statistics
that allow for this interdependency, and again we will turn to randomization
procedures (see chapter 5) involving permutations of rows or columns in the
matrices.

The Mantel Test

Perhaps the most familiar and widely available permutation test for comparing
matrices is the Mantel test (Mantel 1967). The Mantel statistic Z is calculated
as the sum of the products of the corresponding elements of two matrices **X**
and **Y**:

$$Z = \sum_{i=1}^{n} \sum_{j=1}^{n} X_{ij} Y_{ij},$$

where n is, as usual, the number of animals in each network. In the simple $n = 3$ example depicted in table 7.1, we have, running from top left to bottom right of the matrices,

$$Z = (0 \times 0) + (4 \times 1) + (5 \times 2) + (4 \times 1) + (0 \times 0) + (6 \times 3) + (5 \times 2)$$
$$+ (6 \times 3) + (0 \times 0) = 64.$$

The statistical significance of Z is calculated using a randomization strategy in which the rows and columns of one of the matrices are subject to a series of random permutations. In each one, the order in which the animals appear in **X** (say) are scrambled, though not the value of the number relating any two individuals. The test statistic Z is calculated after each permutation, correlating the scrambled matrix **X** with the unscrambled **Y**. The probability value of Z is determined as the proportion of random permutations that result in a higher test statistic than the one calculated for the real data. The Mantel test can be performed with relative ease in a number of statistical packages such as SOCPROG (see box 1.1). A program we have found useful for comparing matrices and very easy to use is Charlotte Hemelrijk's MatrixtesterPrj (see Hemelrijk 1990a; Hemelrijk 1990b), which includes more subtle tests, some of which we will introduce shortly. For the simple matrices in table 7.1, the Mantel test yields, not surprisingly, a perfect correlation, with a correlation coefficient of 1 and $P < 0.001$.

An Application of the Mantel Test

Let us pause to give an example of the straightforward use of the Mantel test to compare the structure of two networks measured on the same set of individuals. In our own work on guppies (*P. reticulata*) we have used the method to investigate whether individuals associate socially with others with whom they engage in cooperative interactions (Croft et al. 2006). We investigated cooperation in the context of predator inspection, a behavior in which individuals leave the relative safety of a group to approach and inspect a predator, gaining information on the predator's state and on the probability of attack (Pitcher, Green, and Magurran 1986). Previous studies suggest that individual fish share the risk by inspecting the predator in a cooperative partnership (Milinski 1987; Dugatkin 1988). To test the prediction that individuals will form stable social associations with animals with whom they cooperate, we compared the association matrix of shoaling behavior (**X**), indicating "who shoals with whom," with associations during predator inspection (**Y**), indicating "who inspects or cooperates with whom," using a Mantel test. The observations used to compile the two networks (shoaling and inspection) were made at different times. To allow us to generalize about the observed patterns, we replicated our study using pairs of networks derived from 19 independent samples. The matrices **X** and **Y** for one of the samples are shown in table 7.2; they each represent undirected relations, so there is no need to show both halves of each matrix; instead the lower left half shows the entries in **X** and the upper right half the entries in **Y**. Each entry in **X** represents the association strength (*AS*) of two fish, and each in **Y** the "inspection strength" (*IS*)—see Croft et al. (2006) for details. For this network, there are 12 fish, $Z = 4,523$, the correlation coefficient $= 0.32$, and $P = 0.019$.

To investigate the overall trend across networks, individual P-values for each network were combined using Fisher's omnibus test (Haccou and Meelis 1992). A significant positive correlation was found between the *AS* of a pair-wise interaction in the social network and the *IS* during predator inspection (Fisher's omnibus test $f_{38} = 77.12$, $P < 0.001$). These results suggest that repeated interactions are a factor in the formation of cooperative interactions during predator inspection in a natural fish population.

A Note on Replication

When we make comparisons between networks, it obviously helps if we can replicate our study to be able to draw more general conclusions. For instance, if we are interested in the friendship network of school children, then it is unlikely that the friendship network of a particular school will tell us all we need to know, and we should try to analyze the friendship networks of a representative sample of schools. The analysis of each network is then followed by a comparison of the networks, from which we may be able to draw general conclusions regarding friendships.

The above logic may seem obvious, yet, as we have mentioned a couple of times before, replication is conspicuously absent in many network studies in the literature. This is particularly worrying when comparisons between different species are being made. If only one network is available per species (from one population, for example) then we do not know whether within-species variation in networks might be just as great as (or indeed, greater than) between-species variation.

More Uses, and Extensions, of Matrix Correlation Tests

With a few modifications the Mantel test can be used to do more than we might at first expect. However, it does have limitations, so it may be worth considering some of the variations on a theme, as we do now. We will follow two rather careful papers by Charlotte Hemelrijk (Hemelrijk 1990a; Hemelrijk 1990b) that confront many issues that arise in comparing matrices of the same size. Her particular interest was to be able to identify reciprocity (exchange of

TABLE 7.2.
An example pair of association matrices from the study of Croft et al. (2006) for a sample of 12 fish (coded A–L). Numbers below the diagonal line represent the association strength (*AS*) between pairs of animals (measured as the number of times a pair of fish was observed in the same shoal over 30 observations made at 1-minute intervals), and those above the line represent inspection strength (*IS*) (measured as the number of times a pair of fish was observed to inspect a predator together over a 30-minute period). For example, animals *B* and *C* have $AS = 20$, $IS = 10$.

	A	B	C	D	E	F	G	H	I	J	K	L
A		6	6	7	5	4	5	6	7	7	7	5
B	17		10	7	6	8	6	6	5	6	4	5
C	21	20		7	7	8	6	8	6	6	4	5
D	16	16	14		6	6	7	7	7	7	4	5
E	12	12	13	15		6	8	11	5	7	5	5
F	11	13	12	16	21		6	8	7	6	3	5
G	10	8	8	9	12	7		7	5	9	4	5
H	8	6	5	12	14	14	7		9	7	5	5
I	14	14	14	13	11	12	12	13		6	5	5
J	10	7	7	7	9	8	12	12	6		5	6
K	10	9	9	10	10	10	11	14	10	14		4
L	6	7	7	8	9	7	8	9	6	10	8	

similar acts) and interchange (exchange of one act in response to another) among primates.

The Mantel test is rather susceptible to outliers, as these will effect all the rows and columns in which they appear. This led Hemelrijk (1990a) to advocate a systematic search not just for "absolute" correlation between two weighted matrices **X** and **Y**, but also "relative" correlation (between versions of the matrices with elements ranked by their weight) and "qualitative" correlation (between binary versions of the matrices). To achieve this, she used other statistics, such as *R*, which is just the Mantel *Z* computed on ranks of the elements of **X** and **Y**, and *K*, which is the matrix equivalent of the Kendall τ (a rank correlation coefficient; see Hemelrijk [1990a] for details), and variations on these measures.

Let us give an example. For a directed network, the element X_{ij} indicates the strength of the interaction from animal *i* to animal *j*, whereas X_{ji} is the strength of interaction from *j* to *i*. Thus, to test whether interactions tend to be reciprocated, we can construct a matrix **Y** whose rows are the columns of **X**

TABLE 7.3.
Matrix of agonistic interventions among captive male chimpanzees.
The matrix is redrawn from Hemelrijk (1990a), which also contains
details of how the strengths of pair-wise relations were measured.

					Receiver			
		Y	D	WO	JO	FO	TA	JA
	Y	0	1	0	0	3	0	2
	D	0	0	0	0	0	0	1
	WO	0	3	0	0	2	2	2
Actor	JO	0	1	0	0	2	1	3
	FO	2	4	0	2	0	0	0
	TA	0	1	4	3	0	0	1
	JA	1	1	0	3	2	2	0

and vice versa (**Y** is the "transpose matrix" of **X**), and perform a matrix correlation test on **X** (sometimes called the "actor matrix") and **Y** (the "receiver matrix"). If there is a significant positive correlation, then interactions tend to be reciprocated.

Hemelrijk (1990a) used this approach to investigate the reciprocal nature of support during aggressive interactions among male chimpanzees at the Burgers Zoo in Arnhem. In this context, support is defined as an agonistic intervention by individual A in a conflict between B and C, which is directed at either B or C but not both. When A directs its intervention at C, then A is said to support B. The starting point is the matrix of support, shown in table 7.3.

Hemelrijk (1990a) first looked for "relative" reciprocity, using a variation of the K test in which the elements in each matrix are ranked within each row (table 7.4). This yields a statistic K_r, which is again tested with a randomization. The result was $K_r = 14$, $P = 0.05$, suggesting that support is reciprocated in a relative way. However, when she compared table 7.3 with its transposed matrix, she found a non-significant "absolute" correlation between the two matrices (Mantel test: $N = 7$, $Z = 70$, $P = 0.12$). We must then ask why there is a discrepancy between the relative and absolute tests? The answer is probably due to the sensitivity of the Mantel test to outliers. We see in table 7.3 that male D did not reciprocate the support that it was given. Removing this individual from the analysis resulted in significant reciprocity, both in relative ($N = 6$, $K_r = 23$, $P = 0.03$) and absolute (Mantel test: $N = 6$, $Z = 68$, $P = 0.04$) terms (results taken from Hemelrijk [1990a]). Hemelrijk notes that male D appeared to be the highest ranking male in this study. If we plot the chimpanzee support network (figure 7.1), this male occupies a central position

TABLE 7.4.

(a) Ranked-within-rows matrix of agonistic interventions among captive male chimpanzees (see table 7.3). (b) Also shown is the transposed form of the matrix, also ranked within rows. Data taken from Hemelrijk (1990a).

		Receiver						
		Y	D	WO	JO	FO	TA	JA
	Y	0	4	2	2	6	2	5
	D	3	0	3	3	3	3	6
Actor	WO	1.5	6	0	1.5	4	4	4
	JO	1.5	3.5	1.5	0	5	3.5	6
	FO	4.5	6	2	4.5	0	2	2
	TA	1.5	3.5	6	5	1.5	0	3.5
	JA	2.5	2.5	1	6	4.5	4.5	0

		Actor						
		Y	D	WO	JO	FO	TA	JA
	Y	0	2.5	2.5	2.5	6	2.5	5
	D	2.5	0	5	2.5	6	2.5	2.5
Receiver	WO	3	3	0	3	3	6	3
	JO	2	2	2	0	4	5.5	5.5
	FO	6	1.5	4	4	0	1.5	4
	TA	2	2	5.5	4	2	0	5.5
	JA	4.5	2.5	4.5	6	1	2.5	0

in the network, at least according to a spring-embedding layout. Perhaps male D simply does not need to support other males very often.

Permutation tests like these can also be used to test models of network structure by constructing a so-called "hypothesis matrix" (Hemelrijk 1990a). In this case, **X** is again an empirical association matrix constructed for a particular type of interaction. **Y** is a matrix constructed by the researcher to reflect a trait (or attribute) within the population that he or she hypothesizes might be responsible for the network structure. The hypothesis matrix is constructed on the basis on some measure of similarity between individuals such as relative dominance, relatedness, or difference in body size (Schnell, Watt, and Douglas 1985). Note that UCINET has a helpful capability for constructing a hypothesis matrix from node attribute data, under the *data > attribute* menu. This capability requires a list of node attributes as input, from which it will

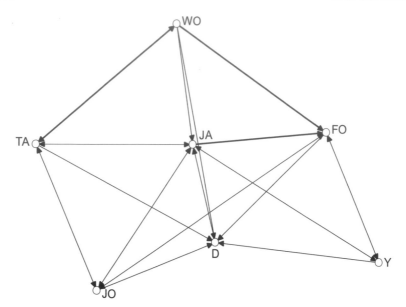

Figure 7.1. The network of agonistic interventions among male chimpanzees, drawn from table 7.2.

construct the matrix of differences between individual values, among various choices.

Hemelrijk (1990a) used the hypothesis matrix approach to test on female chimpanzees a hypothesis of Seyfarth (1976, 1980). He suggested that, provided there is little competition for a grooming partner, both baboons (*Papio cynocephalus ursinus*) and vervet monkeys (*Cercopithecus aethiops*) groom a female more often if she has a high dominance rank. Hemelrijk (1990a) constructed an association matrix (**X**) of "who grooms whom" among eight female chimpanzees, and another matrix (**Y**) that reflected the difference in dominance rank between pairs of females (with higher numbers representing a more dominant rank). She found significant relative correlation between her matrices ($N = 8$, $K_r = 53$, $P = 0.0005$), suggesting that females do indeed tend to groom dominant females more often.

In a second paper, (1990b) Hemelrijk extends the hypothesis-matrix approach to control for a third variable when comparing two networks. Of course, any test of correlation between two variables cannot tell us whether there is a causative relation between them, but it can be useful to consider the role of other factors. Hemelrijk's motivation can be illustrated by reference to the last two examples; if primates groom female partners more often if they have high rank, then it is possible that the correlation between grooming and the receipt of support in conflicts may be spurious, and merely the by-product of the correlation between grooming and dominance. To explore this possibility, Hemelrijk (1990b) revisited the data of Seyfarth (1976, 1980) on grooming, support, and dominance in baboons and vervet monkeys. She developed and applied a partial-correlation version of her K_r test, constructed for matrices by analogy with the Kendall partial rank correlation coefficient (Kendall 1962). Her analysis suggested that the interchange of grooming for the receipt of intervention during fights is indeed a spurious correlation in the vervet monkey data, but not in the baboon data.

A number of programs will calculate partial correlations with relative ease, such as MatrixtesterPrj. For a detailed discussion of these, and the other methods discussed here, see the excellent papers by Hemelrijk (1990a, 1990b). These papers contain other variants on matrix correlation tests that are beyond the scope of this chapter. An important message to take from these papers is that care must be taken when searching for correlations between relational data sets. Some of the more advanced tests, such as the K_r test, appear to overcome problems with statistical outliers, but even with the best choice of test, it is essential that any analysis be careful and thorough.

7.2 COMPARING NETWORKS FOR DIFFERENT INDIVIDUALS

As we have already hinted, things get a little trickier if we wish to compare two or more networks constructed for different individuals, from different populations, or different species, though there are plenty of reasons to want to make such comparisons. These difficulties must be overcome not just if we want to compare the social structures of species A, B, and C, for example, but also if we want to use networks to analyze long-term data sets on the same population in some sort of longitudinal network analysis. If we are lucky enough to have been funded sufficiently well to have looked at the same system more than once, or better still for longer than the typical lifetime of our study species, then it is a tantalizing prospect to slice our data into appropriate periods of time, construct a social network for each slice, and trace the structural evolution of the networks over time. Typical time slices might be based on short periods for a lab-based study or, in the field, repeated samples of a population in successive years. Few examples so far exist, though we are sure that plenty of appropriate long-term data sets exist that would be amenable to such analysis.

In one example we are aware of, Cross et al. (2004) constructed association-based social networks for a population of African buffaloes (*S. caffer*) in 2002 and 2003. For consistency, they included in their analysis only the 64 radio-collared buffalo that survived both years. The principal aim of this study was to look at the potential effect of social network structure on disease dynamics. From the social network for each year, they conducted a hierarchical cluster analysis (see chapter 6) to build dendrograms of associations among the 64 animals. They compared the networks constructed in 2002 and 2003 by making largely qualitative comparisons between the dendrograms. They found that in 2002, the buffalo formed three rather tight clusters, which became less well defined in 2003, with more associations occurring between clusters (and hence a greater chance for some diseases to spread through a larger fraction of the population). The authors suggest that the loosening of the hierarchical social structure may have been due to large movements of the animals in response to particularly dry conditions at the end of 2002.

If it turns out that more general network comparisons, where the individuals are not the same in each network, can be made robust, then the possibilities of what might be explored by comparing networks are mouth watering, both for longitudinal analyses of single (dynamically changing) populations and also between populations and species. However, there are issues that potentially get in the way. The first, and perhaps most important, is that we must be able to compare two or more networks of different size, containing a different number of individuals. For a start, this means that we cannot directly compare the network matrices using the Mantel or any similar test. We might then decide to concentrate on comparisons between network measures, such as the degree, the path length, or the clustering coefficient (see chapters 4 and 5).

As an example of this approach, Barabási et al. (2002) looked at the evolution of the network of scientific collaborations (who published with whom) between 1991 and 1998 by comparing the structure of networks generated from time slices of one year. The authors compare node-based measures including the mean degree and clustering coefficient. They conclude that "preferential attachment" (the tendency of new authors to publish with established, well-connected authors) is the driving mechanism in the evolution of this network. This is an interesting study, but crucially it depends on the use of a model of network evolution. This is needed, at least in part, to overcome some of the pitfalls of directly comparing structural metrics derived from networks of different size and number of edges. We will see that developments of longitudinal (and other) network comparisons in the social sciences have also come to rely heavily on the use of network models. To see why we can't just blithely compare network metrics, we need to take an interlude.

Effects of the Numbers of Nodes and Edges

The issue here is simply that most network measures vary with the number of nodes (n) and distinct edges (E), so we must try, if possible, to control for this. The point can be illustrated with a simple example. Suppose that we construct networks in three successive years for a growing population, with $n = 100$, 200, and 300 in each of the years. For the sake of argument, suppose that each year we find that the mean degree is $k = 10$. (This in itself might be the most

TABLE 7.5.
Some hypothetical values of mean path
length (L) and clustering coefficient
(C) from networks constructed from a
growing population of size n.

	Year		
	1	2	3
n	100	200	300
L	2.0	2.3	2.5
C	0.1	0.05	0.03

interesting result from our study, but we digress.) Then suppose we decide to
use the mean path length, L, and clustering coefficient, C, to characterize the
networks, and we find the results shown in table 7.5. How might we interpret
these results, comparing structural measures of an evolving social network,
in terms of the evolution of social structure in our population? Is the society
becoming less cliquey (C reducing) and more "strung out" (L increasing) as
time passes?

Well, perhaps not. We have deliberately chosen some figures here that are
completely consistent with those we would expect from an Erdös-Rényi random network in which each edge is placed between two randomly chosen
nodes. As we saw in chapter 4, such a network has, for a constant mean degree, a mean path length that increases logarithmically with n (equation 4.8)
and a mean clustering coefficient inversely proportional to n, exactly as we
see in table 7.5. In other words, a "growing" network with completely random
interactions would be changing with time in exactly this way! Other similarly
spurious variations in network measure can easily be produced, as they also
depend strongly on the number of edges (E).

So unless we can orchestrate our studies so that all networks we wish to
compare contain the same number of nodes and edges, direct comparison of
simple node-based measures is a dangerous business. If we do wish to make
such comparisons, it would be wise at least to replicate the calculations of
table 7.5 to give some sort of baseline of how much difference to expect in
node-based measures. If n and E can't be matched, the next best thing is probably to try to compare networks with a similar edge density (section 4.1). We
should note, of course, that incomplete sampling will lead to a misrepresentation of both n and E in any one network, and unless we can be confident that
we have a similar sampling error each time we construct a network, any comparison between them is again subject to these problems.

Of course, if we do have a situation in which the number of nodes and edges
are fairly well matched, none of this is so much of a worry. Sundaresan et al.

(2007) constructed networks using both association strength and the half-weight index (see chapter 3) for 28 Grevy's zebra (*Equus grevyi*) and 29 onager (*Equus hemionus khur*). These closely related equids were previously thought to have the same social structure, despite living in rather different environments. Sundaresen et al. (2007) compared the clustering coefficient and path length (among other things) between the two networks, using two-sample permutation tests (Good 2000), a randomization technique. They were able to conclude that the social structure was not the same in the two species, with differences consistent with the environmental conditions. For example, Onager live in open desert, so can more easily re-form associations than can Grevy's zebra that live in bushier habitat. This is reflected in the fact that the Grevy's zebra networks have a significantly larger clustering coefficient, indicating greater cliquishness.

The analysis of Sundaresen et al. (2007) contains methodological steps that we have identified earlier as potentially problematic, such as the computation of dyadic P-values to construct networks (see chapter 5). Nonetheless, the whole approach of using networks to make a quantitative comparison of social structure is exactly what we would like to achieve. So how do we proceed if, unlike the authors here, we have networks of rather different size or density? Let us consider first some of the more straightforward methods by which we might help ourselves. If we are interested in comparing node values for individuals, or groups or classes of individuals, between networks, then it might be wise to consider ranks of values, rather than the values themselves. If young males always occupy the highest ranks of node betweenness, it would seem reasonable to base some test on this, rather than on the actual values of betweenness, which vary considerably with n and E.

A different approach would be to rescale our network measures by dividing them by some fixed values. This is a familiar tactic in many comparative measures. The coefficient of variation, the opportunity for sexual selection (Arnold and Wade 1984), and diversity indices (Magurran 2003) are all examples of dimensionless parameters that permit comparison between different regimes. The situation is not quite the same here, as network measures are just numbers, and are therefore already dimensionless, but some rescaling of the range on their values might still be useful. The next step is to decide what values to use to rescale network measures. One ploy would be to scale them all to lie between zero and one by dividing them by their maximum value ($n - 1$ for node degrees, the network diameter D for the path lengths, and so on). While this might help with some comparisons between networks, it would not have revealed that the data in table 7.5 are all consistent with a simple random network of interactions.

Watts and Strogatz (1998) took another approach. They scaled path lengths and clustering coefficients to the values expected for an Erdös-Rényi random network with the same numbers of nodes and edges in order to show that many diverse real-world networks exhibit "small-worlds" properties. This choice

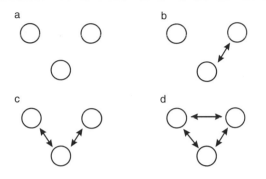

Figure 7.2. The four possible types of triadic relation in an undirected network: (a) no edges, (b) one edge, (c) two edges, and (d) all three edges.

would certainly have helped with our fictitious case, and would have revealed that we had entirely consistent random networks in each year. But this is not the only choice of scaling values; there is no particular reason to take a particular, or any, random network as our baseline. There are many other networks we could choose, and each might have merit; if we know there is a strong substructure in some or all of the networks we are comparing (see chapter 6), caused for example by spatial fidelity of our animals, then it could be argued that measures should be scaled in some way to reflect this. Of course, this is the very point we want to make here: we are led to the view that what we should do is compare our structures with some sort of model of network formation that controls for features such as the number of nodes and edges, or gross structural features, and see whether some networks are more consistent with our model than others.

Triads and Other Structural Motifs

As a precursor to our discussion of models used in the social sciences, we should highlight a different approach to characterizing network structure that is based not on the evaluation of node-based metrics such as those we used in chapters 4 and 5 but on performing a "census" of small sub-networks ("motifs") that we hope exhibit key local structure. These motifs are often "triads," or combinations of edges between three nodes (Wasserman and Faust 1994). For an undirected network, there are four geometrically distinct ways in which three nodes may be connected by edges, shown in figure 7.2.

The idea is that we can examine the structure of an undirected network by counting the frequency of the four types of triad relations across all possible triplets of nodes. This is a "triad census," which will provide quantitative information on the social structure of the population; in particular it will tell us the frequency of isolated individuals, couples, structural holes—where one actor

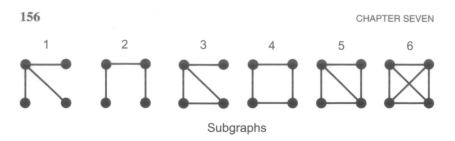

Subgraphs

Figure 7.3. An alternative set of motifs for structural comparison of undirected networks, after Milo et al. (2004).

is connected to two others who themselves are not connected (figure 7.2c)—and clusters of individuals (figure 7.2d). The idea is to compare the "spectrum" of frequencies between networks as a means of comparing the networks themselves. Network software such as Pajek (available as part of UCINET, for example) includes the option of a triad census (de Nooy, Mrvar, and Batagelj 2005).

Two things may have occurred to the vigilant reader. First, there is no null model yet, so no way of validating any comparison. This will come shortly. Second, the triads in figure 7.2 don't appear to tell us much that we couldn't have learned from simple node-based measures. After all, the clustering coefficient is just a measure of the ratio of "all edges" triads to "two edges" triads in an undirected network, and the node degree will give us a good idea of how many edges there are per triad. In an alternative approach to much the same problem, Milo et al. (2004) used the six motifs shown in figure 7.3, each of which contains four nodes, to compare the structure of various undirected physical and biological networks. The relative frequency of each motif was found to be very similar within families of contexts (such as networks of proteins and autonomous systems), though different between contexts. In each case the authors suggested model networks that have similar motif spectra, and so might serve as reasonable models of network structure.

The idea of a triad census really comes into its own (as does most modeling in the social sciences) when the network is directed. Then it is much harder to find a simple metric that measures local cliquishness, and a triad census is a more appealing approach. There are sixteen possible types of triad in a directed network, shown in figure 7.4. These motifs, or subsets of them, have been used as a basis for various network comparisons. Milo et al. (2004) used a triad census to classify a wide range of large directed networks. They compared the frequency of occurrence of the motifs with randomized versions of the networks in which the number of nodes, edges, and the degree of each node was conserved. The result of all this was a "significance profile." They applied their method to several biological, technological, and sociological networks, and found several "superfamilies" of networks not previously thought to be related but which shared similar significance profiles. Though the structural

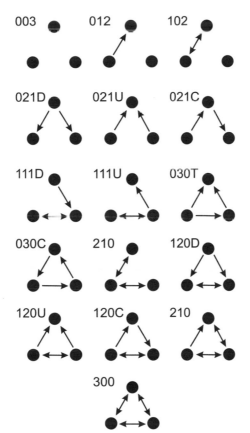

Figure 7.4. The 16 possible types of triadic relations for directed data. In the labeling system the first character gives the number of mutual dyads in the triad. The second character gives the number of asymmetric dyads in the triad. The third character gives the number of null dyads in the triad. Lastly, the fourth character is used to distinguish the triad among other types: T = the triad contains transitivity, C = the triad is cyclic, D = down, U = up.

signatures are different, and the nature of the randomization tests not quite the same, the type of analysis in Milo et al. (2004) is not far in spirit from the node-based analysis presented in chapter 5.

In a similar vein, Freeman (1992) studied group structure and social interactions in humans via a census of "transitive" triads (those in which individual A is connected directly to B and also via C). He constructed networks in different settings, including interactions during lunch, at the beach, in the office, and at a karate club. Freeman counted the number of intransitive triples in the

networks and compared these to those predicted by two different models of
how humans form social groups.

7.3 COMPARING NETWORKS VIA STATISTICAL MODELS

For our final section, we will go some way out of our comfort zone to try and
describe some of the methods that have been developed in the social sciences
for the characterization, and subsequent comparison, of network structure.
We feel that we should apologize in advance to professionals in this field for
any misrepresentation of their work. The methods we are interested in have
made use of a family of statistical models originally designed to characterize a single network by fitting coefficients of model parameters representing
node attributes ("explanatory actor variables" in the language of these models)
or relational attributes at the level of the edge (dyad) and triad, such as those
depicted in figure 7.4. The models are all variations on a theme of generalized linear models (Dobson 2001). In the vast majority of cases, the networks
modeled (and then compared) this way have been directed, though there is no
obvious reason why this should be a requirement. It is, though, central to the
particular models presented in this section that the networks are unweighted,
so that each edge may be treated as a "dichotomous" variable: either it is there
or it is not.

Logistic Regression, Logits and p* Models

The network comparison techniques we will outline all make use of logistic
regression (Kleinbaum 1994), a well-known method to analyze dichotomous
dependent variables. If the binary dependent variable in a "conventional" logistic regression is X (either 0 or 1), p is the probability that $X = 1$, and there
are r explanatory variables, denoted $z_1, z_2, \ldots z_r$, then p is related to z_i via the
logarithm of the odds (or "logit") of p, given by

$$\mathrm{logit}(p) = \ln\left(\frac{p}{1-p}\right) = \alpha + \sum_{i=1}^{r} \beta_i z_i.$$

The result of the logistic regression is the evaluation of the constant α and
the coefficients β_i that best fit the observed probabilities p. These coefficients
might be estimated by a maximum likelihood method (Sokal and Rohlf 1994).
Often the aim of a logistic regression is to search for the most parsimonious
set of explanatory variables that are consistent with the observed probabilities
p. The relative importance of the various explanatory variables may be explored using a Wald test, a likelihood ratio test, or something similar (Sokal
and Rohlf 1994).

The possibility of using an approach based on logistic regression to characterize the structure of a social network has been recognized for some time (Holland and Leinhardt 1981). In principle, each of the elements X_{ij} of an unweighted association matrix can be treated as a dichotomous variable, and the probability that each edge exists can be related through a logit function to a set of explanatory variables, which might include node attributes or simple motif-based relational attributes of the types we have introduced in the previous section.

There is, however, one rather major snag to all this. One of the assumptions of a logistic regression is that the data points are independent, and this is certainly not the case if the data points are the elements of an association matrix; we have already mentioned several times that elements in a matrix are not independent. The way around this is rather subtle, and the details are well beyond the scope of this book. The interested reader should consult the chapter by Wasserman and Robins in Carrington, Scott, and Wasserman (2005), and references therein, for an account of what may be done to allow for certain dependencies among the elements of an association matrix. The upshot of all this is that there are families of dependencies that can apparently be dealt with, and these lead to families of models, all based on logistic regression. One of the more favored families of models are the so-called p* models (Freeman 1992; Wasserman and Pattison 1996; Pattison and Wasserman 1999), in which any two edges that share a node are considered to be dependent, and this dependency is taken into account.

So the essence of a p* model is that we have a logistic regression in a series of explanatory variables. If we buy into the fact that awkward issues regarding data dependency can be side stepped, the good news is that these explanatory variables need not be restricted to node attributes (sex, size, and so on) but can also reflect relational properties such as a tendency for edges to be reciprocated (so if A is connected to B, B is also connected to A in a directed network) and even relational properties based on interrelations between three nodes. In other words, p* models are able to incorporate the effect of the prevalence of the triads in figure 7.4, as well as dyadic and node-based features.

7.4 COMPARISONS ACROSS POPULATIONS AND SPECIES

An example of using the p* models to make comparisons between network structures is the work by Faust and Skvoretz (2002), who examined the structural features of 42 directed networks from three types of species (humans, nonhuman primates, and non-primate mammals). The networks varied in size between 4 and 73 individuals. The relations in the networks covered a range of interactions from licking and grooming to disliking and victory in agonistic encounters. The authors were interested to see whether structural patterning in

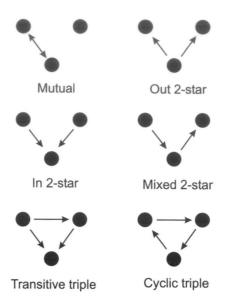

Mutual   Out 2-star

In 2-star   Mixed 2-star

Transitive triple   Cyclic triple

Figure 7.5. The six dyad and triad structural elements used by Faust and Skvoretz (2002).

a network could best be predicted by the type of animal or by the definition of an interaction. The expectation was that networks with similar social relations would have similar social structure.

As a starting point for their comparisons, Faust & Skvoretz (2002) fitted a p* model to each of the 42 networks separately, using the same set of six structural explanatory variables shown in figure 7.5 in every case. Each network was then characterized by six coefficients from the p* model. Rather than use these coefficients to compare the networks, the authors used the coefficients for one focal network to compute the probability that each edge in that network, and each of the other 41 networks, should exist, were the p* model a perfect representation of the structure of the focal network. These pair-wise comparisons were then used to generate a dissimilarity score between the predicted edge probabilities for the focal network and for each of the other networks. This process was repeated by trying each of the 42 sets of coefficients on every edge in each of the 42 networks. Finally, a correspondence analysis (Greenacre 1984) was used on the 42×42 matrix of dissimilarity scores to compare all the networks and to relate the comparisons to the model inputs.

Faust and Skvoretz (2002) found some interesting differences in network structure between species and contexts (though note that there might be caveats to the approach, as we mention in our concluding remarks at the end of the chapter). They suggested that networks derived from a given context may have common structural features. For example, networks constructed for fighting

and dominance were different from networks constructed for relations of af-
fection and affiliation. There were also differences observed between species;
humans for example, showed more of a tendency for mutualism (reciprocated
edges) than nonhuman primates. More amusingly, the authors suggest, as part
of an illustration of pair-wise structural comparisons, that the model of social
licking among cows is the best predictor of co-sponsorship among a group of
U.S. senators.

7.5 LONGITUDINAL ANALYSIS IN THE SOCIAL SCIENCES

Models following what appear to us to be a similar theme have been used by
Tom Snijders, in particular, to facilitate longitudinal analysis of (human) social networks. His analyses (Snijders 1996a; Snijders 1996b) are conducted on
series of networks, representing more or less the same set of individuals. The
basic idea is to use the first observed network as a given, then use a dynamical
model to fit the subsequent observed networks. The method makes the most
of having more information on social structure than could be gleaned from a
single snapshot. The models investigate the key elements in network evolution
from one observed network to the next as a function of explanatory variables,
which again may be node-based or reflect local structural features associated
with dyads or triads. The dynamics are modeled by assuming there are many
incremental time-steps between each observation. Each node is treated as a
purposeful agent that acts myopically (considering only its immediate neighbors at each instant of time) to optimize its "objective function," but subject
to some random variability. The objective function is parameterized in terms
of the explanatory variables. At each time-step, one node is chosen at random, and potentially switches one of its edges off or makes a new edge to
another node. The probability that a node makes a change is determined from
a logit function of the explanatory variables (and therefore depends on what
the change would do to the node's objective function), so in structure at least,
these models are rather similar to others in this section.

One of the extensions to longitudinal models looks particularly interesting.
"Actor-driven" models are a means to integrate the simultaneous evolution of
networks and behavior. At each time-step an actor controls not only his outgo-
ing edges, but also his behavior, so the structure and behavior are able to influence one another. As an example, in an investigation of the effects of delinquent behavior on friendship formation, Snijders and Baerveldt (2003) studied
the evolution of friendship networks in nineteen school classes, and concluded
that there was evidence for an effect of similarity in delinquent behavior on
friendship evolution, with the degree of delinquent behavior having a positive
effect on tie formation but also on tie dissolution. This sort of insight into the
factors important in determining the organization and development of human

social networks is an approach that may find use in study of animal behavior. Models for studying the continuous evolution of networks and the methods for implementing them are reviewed by Snijders (2005).

7.6 CONCLUDING REMARKS

Before we leave these models, we should sound a note of caution. A great deal of the detail of how and why the models presented here work has been omitted, and it would be unwise to launch oneself into any similar analysis without a good deal of thought and preparation. It would also appear that there may still be unresolved issues with the methodology. Faust (2006) has investigated the effects of network size (n) and density (ρ) on the observed frequency of triads across diverse networks including social networks from humans, nonhuman primates, and nonprimate mammals. Her findings suggest that similarities in censuses of triad motifs across these networks can be largely accounted for by dyadic and node-based features of the networks. Thus the desired effect of avoiding issues to do with the number of nodes and edges may not have been achieved. On a similar theme, Solé and Valverde (2006) have suggested (in the context of cellular networks) that motif structures might merely be "spandrels" or, for those of you without a dictionary to hand, by-products of inevitable rules of network structure, and of no intrinsic interest.

Whatever the immediate uncertainties or difficulties, the prize of incorporating network structure and evolution into the framework of the comparative method is surely worth pursuing. It seems likely that methods for comparing networks will involve comparisons mediated by models, whether they be a means to parameterize the structure or null models aimed to control for the key biological, ecological, spatial, or temporal aspects of the social system under investigation. Developing methods for comparing animal social networks of different sizes and densities is going to be a challenge for future research.

8

Conclusions

Social network theory provides a general framework that allows us to visualize and quantify global and fine-scale social organization. With a bit of imagination there are a number of ways that social networks can be constructed (see chapter 1). This book has largely focused on social networks in which nodes represent individuals, and we have paid particular attention to issues that arise when association data are constructed via the "gambit of the group" (see chapter 2). Visualization of the network is only the starting point for detailed analysis (chapter 3), but its importance should not be underestimated. Looking at the social network of your study species for the first time if often an exciting and stimulating experience.

At the heart of the networks approach is a powerful set of novel quantitative methods and measures that can describe the patterns of social organization at many levels, from individuals to populations. One of the features that makes a networks approach distinctive is that it pulls together ideas and methods from a wide range of disciplines. In chapters 4–7 we have invoked methods and ideas from social scientists, biologists, mathematicians, physicists, computer scientists, and others. This broad interest is incredibly productive, and will almost certainly continue to provide novel approaches that we can harness, provided we remember to apply these ideas sensibly.

At the level of the individual we may calculate descriptive statistics such as an individual's degree or betweenness (chapter 4). Such measures can be averaged over the network as a whole and used to describe social organization at the level of the population. As with all quantitative variables, it is important to establish whether the measured values differ from those expected by chance. We can determine the significance of measured network metrics by comparing a measured value to that expected by random interactions or to a constrained null model where we preserve certain features of the data, thus adding important statistical rigor to network studies (chapter 5). Integration of these quantitative network metrics with traditional ecological and evolutionary studies is expected greatly to advance our understanding of the mechanisms, development, evolution, and functions of social behavior.

A networks approach also allows us to examine intermediate patterns of social organization between that of the individual and the population. For example, we may search the network for communities, defined in direct analogy to human social communities as collections of individuals that are more connected amongst themselves than with other individuals in the network (see chapter 6). Understanding the structure of these communities may be of particular importance for understanding the roles of individuals in population structure. Work by Lusseau and Newman (2004) on bottlenose dolphins (*T. truncatus*) has shown that dolphin networks exhibited substructures that fulfill the definition of communities, and are interconnected by individuals of high betweenness. Lusseau and Newman suggest that these individuals are "brokers" that may play a crucial role in maintaining the cohesiveness of the dolphin population. Although this hypothesis is plausible, one of the shortfalls with many descriptive network studies is that hypotheses remain essentially untestable because experimental removal is not an option. Such manipulations may be possible for populations that can be observed under laboratory conditions (see Flack et al. [2006] for a very good example).

An area of great interest in behavioral ecology is the issue of how ecological conditions shape the behavior of individuals and thereby the social organization of a population or species (Harvey and Pagel 1991). Closely related species (or different populations of the same species) are compared to investigate how differences in ecological conditions may have led to the evolution of differences in behavior and morphology. The construction of social networks allows us to obtain information about the social fine structure of populations, and we can use this information to answer the question of how differences in ecological conditions have shaped the interaction patterns of individuals. The strength of the social network approach in this context is that it gives us a wealth of statistical descriptors that characterize the social fine structure of an entire population, ranging from measures that focus on individual connectedness in the network to community structures that form the building blocks of a network to global descriptors of the entire network. While there are a number of outstanding issues that need to be addressed in this area (see chapter 7), the comparison of networks between species and between contexts is an exciting avenue for future research, and we expect the rewards for solving the analytical problems to be substantial.

8.1 PREVIOUS APPLICATIONS

As we have tried to make clear throughout the book, the application of network theory to the study of animal populations is still relatively new, but not entirely so. In table 8.1 we give a list of some of the empirical work on animal social networks so far. A complete list would be impossible because it remains a little subjective which studies to include and which not to include. We selected some of the more prominent studies from different taxonomic groups to provide a general overview. Some studies constitute borderline uses of network theory, in that they use the term "social network," but don't apply any of the network methodology. Other studies present visualizations of networks as sociograms, but present no statistical analysis of them. We have added an asterisk to those studies where the social network approach plays a central role in connection with detailed statistical tests. Many more studies have analyzed the structure of association matrices in depth without using the term "network" at all; many of these are reviewed by Hal Whitehead (2008).

From table 8.1 it is obvious that most of the social network studies to date have been carried out on nonhuman primates. Given the restriction on space, we have only named a few studies on humans, but there is of course a vast number of papers in the psychology and sociology literature, with at least one journal dedicated specifically to this sort of work in humans. The books we have referred to many times by Scott (2000), Wasserman and Faust (1994), and Carrington, Scott, and Wasserman (2005) serve as an excellent introduction to work on human social networks.

The prevalence of studies on nonhuman primates should not surprise us given that social network theory was developed to investigate human relations, and therefore this approach relatively easily transfers to other primates. At the same time it is clear that many criteria that we expect from scientific research (such as standardization, replication, manipulation, etc.) are more easily fulfilled with some of the other taxonomic groups. The social insects, in particular, stand out as a taxonomic group that should be highly promising for the network approach. The evolution of division of labor and the frequent interactions among workers belonging to different castes cries out for an approach in which system performance can be developed from social subunits and ultimately dyadic social interactions. Fewell (2003) developed a process-oriented network to study division of labor in honeybees (figure 1.3; table 8.1), but there must be possibilities for networks based on individually marked colony members.

Domestic animals (such as cows, pigs, and sheep) are another group of species for which the network approach would appear to have high potential. The fact that these animals can easily be marked individually and observed continuously (often using automated systems) should provide us with a wealth of information on their social relationships to help address some of the important and urgent welfare issues.

8.2 OUTSTANDING ISSUES AND GOOD PRACTICE

One of the key points we have tried to emphasize throughout this book is that, while there is much to be gained from the application of social networks analysis to animals, we need to be armed with an appreciation of potential pitfalls and limitations. In particular we need to be careful when borrowing

TABLE 8.1.
Empirical studies of animal social networks by taxonomic group. Asterisks indicate those studies where the social network approach plays a central role and is supported by detailed statistical tests.

Species	Authors	Research area
Insects		
Social insects	Fewell 2003*	Modulation of foraging behavior
Fish		
Guppy (*Poecilia reticulata*)	Croft et al. 2004a*, 2005*, 2006*	Patterns of social interaction in a wild population Replicated networks Cooperation
Three-spine stickleback (*Gasterosteus aculeatus*)	Croft et al. 2005*	Patterns of social interaction in a wild population
Birds		
Long-tailed manakin (*Chiroxiphia linearis*)	McDonald 2007	Predicting fitness from network connectivity
Cetaceans		
Dolphin (*Tursiops truncatus*)	Lusseau 2003*; Lusseau et al. 2006*	Patterns of social interaction in wild populations
	Lusseau & Newman 2004 *	Which individuals are important in tying a network together
Pinnipeds		
Sea lion (*Zalophus wollebaeki*)	Wolf et al. 2007*	Patterns of social structure
Primates		
Rhesus monkey (*Macaca mulatta*)	Chepko-Sade et al. 1989	Social structure
	Corr 2001	Changes in social networks over time
	Berman et al. 1997	Infant social networks
	Deputte & Quris 1997	Infant gender and socialization process
	de Waal 1996	Development and perpetuation of affiliative networks
Pigtailed macaque (*Macaca nemestrina*)	Flack et al. 2006*; Flack & Krakauer 2006*; Flack & De Waal 2007*	Policing and stabilization of social niches

TABLE 8.1. (*Continued*)

Species	Authors	Research area
Ring-tailed lemur (*Lemur catta*)	Nakamichi & Koyama 2000	Intra-troop affiliative relationships of females
Northern muriqui (*Brachyteles arachnoids hypoxanthus*)	Strier et al. 2002	Male–male relationships
Emperor tamarin (*Sanguinus imperator*)	Knox & Sade 1991	Agonistic networks
Nicaraguan mantled howler monkeys (*Alouatta palliata*)	Bezanson et al. 2002	Social structure
Chimpanzee (*Pan troglodytes*)	Hemelrijk 1990a*; Hemelrijk 1990b*	Reciprocity and interchange
Western lowland gorilla (*Gorilla gorilla gorilla*)	Stoinski et al. 2003	Proximity patterns of female western lowland gorillas
Western gorilla (*G. g. gorilla*)	Bradley et al. 2004	Extra-group, kin-biased behaviors
Spider monkey (*Ateles geoffroyi*)	Pastor-Nieto 2001	Food sharing
Primates (comparative study)	Kudo & Dunbar 2001	Network size and neocortex size
Ungulates		
African buffalo (*Syncerus caffer*)	Cross et al. 2004*, 2005*	Disease transmission, social structure
Grevy's zebra (*Equus grevyi*) and Onager (*Equus hemionus khur*)	Sundaresen et al. 2007*	Comparing social structure
Elephantidae		
African elephant (*Loxodonta africana*)	Wittemyer et al. 2005	Multiple-tier social structure
Marsupials		
Brushtail possum (*Trichosurus vulpecula*)	Corner et al. 2003*	Social-network analysis of disease transmission
Domestic Animals		
Sheep	Webb 2005	Contact structure for disease modeling
Pig (large white landrace)	Durrell et al. 2004	Preferential associations

methods and theories developed in other disciplines and applying them to social networks. Moreover, there is a need for more empirical studies that use the standard techniques of replication, manipulation, and statistical testing of observed patterns.

Sampling Issues

One of the limitations of empirical work on animal social networks is that information on the network structure will be incomplete because of the restrictions on observing animals and the effort required to collect information on association patterns. What proportion of individuals do we need to include in the network to get a representative network? For how long do we need to sample, and how often? While in chapter 2 we suggest some approximate indicators to gauge how likely we are to have a representative sample, there is a need for future research to address these questions empirically. Furthermore, with the continuous development of remote sensing technology and "smart tags," it will soon be economically feasible to deploy tags on some species (this will largely depend on body size) that can keep track of their social contacts, opening up many new possibilities for analyzing animal social networks.

The Importance of Replication and Manipulation

In some cases we have to be careful with the interpretation of social network data. Take the example of two men who buy a newspaper at the same kiosk at about the same time every morning because they take a particular bus or train to work. In a social network that is based purely on spatial association, they would show up with a strong connection, despite the fact that they might be completely oblivious of each other's existence. Although this type of association may be important for some questions that we want to ask, for example regarding airborne disease transmission, for other types of questions we need to be careful with the interpretation of some network connections that are based on proximity alone. That is, co-occurrence can lead to close social relationships, but does not have to. We have to examine our network data closely to find out whether individuals that are connected in our network just happen to have a similar routine or speed of locomotion, or whether their association is based on social preference. One way of finding out is to perform choice or preference tests in which individuals that have been found to co-occur in the wild have to pick each other out from other individuals that are matched for size, sex, and other characteristics. Croft et al. (2006) showed that guppies that frequently co-occurred in the wild also preferentially shoal with each other when size- and sex-matched control individuals are introduced to the same tank in a laboratory setting. Similar choice experiments have been carried out using sheep (Michelena et al. 2005).

Another way of testing network patterns is to use manipulation experiments. Individuals can be removed from (or added to) a population permanently, or removed, subjected to some treatment (such as training for a task or given specific information or infection) then reintroduced. If environmental or habitat features are thought to be driving social structure, these might be manipulated. Of course, in all of this we must have adequate controls for our procedures. An example of a networks-based manipulation experiment is the elegant work by Flack et al. (2006), who studied the social organization of pigtailed macaques (*M. nemestrina*). They employed the network approach to characterize the social organization of a group of monkeys, making full use of several novel statistical descriptors that allow a detailed assessment of the social relationships between individuals. A removal experiment demonstrated that the presence of several high-ranking individuals was essential to retain group cohesion and that the social network fractured into several subunits in their absence due to increased levels of conflict between individuals. Network tools provide sensitive measures of both individual- and group-level characteristics that are particularly useful in this type of setting.

Statistical Issues and Interpretation

Animal social networks have so far often been used only as graphical tools to visualize social structure, often with no statistical analyses of the network patterns, meaning that no firm conclusions can be drawn. As with standard analytical approaches, the observed patterns need to be compared to those expected by chance. Randomization tests are extremely useful in this regard, due to the difficulties in dealing with relational data sets, in which data points are not independent. Tests range from a simple randomization of ties over the network to more complex null models that conserve important characteristics of the data (see chapter 5). As might be expected, some thought must be given to the choice of randomization test used in any particular case. If we believe that we have a truly representative network, simple randomization of node or edge labels may do the trick. As we discussed in chapter 5, there are more unresolved issues if we have used the gambit of the group to construct our network. Then we have to decide how to weight the edges, whether and how to filter the nodes and edges, and whether it should be the structure of the original data set that should be conserved in a randomization test rather than, say, the number of edges.

The issues associated with testing the significance of networks patterns has received relatively little attention in the networks literature, with the notable exception of the social sciences. In chapter 7 we discussed briefly some of the statistical models developed in this field to characterize network structure. Further work exploring the application and extension of such models to animal populations may prove very fruitful.

While borrowing methods, ideas, and concepts from other disciplines may be very productive, we need to proceed with some caution. As we mentioned in chapter 4, there is a danger that we devote too much energy to a search for tags such as "small-worlds" or "scale-free" for our networks, without paying enough attention to whether such descriptions are statistically meaningful for small networks (in the case of scale-free) or almost inevitable (as is the case for some small-worlds properties).

Finally, we need to consider carefully the limitations of our data collection on any interpretation of our network. This is particularly true when it comes to analyzing processes on networks. Our book has been almost entirely on pattern, not process, and though there is every reason to want to extend the use of animal social networks in this direction, we must be careful. Often we will be forced by circumstance to use point sampling to quantify association patterns, making our observations once per day, week, month, or even year. While this may give us a representative impression of certain aspects of the social structure, we cannot infer the dynamics that underpin this structure. This is another example of where we must avoid jumping onto a bandwagon. One of the major successes of recent network theory has been to show that the spread of epidemics is quantitatively dependent on network structure (see Boccaletti et al. [2006] for a review). The analysis is based on infections moving through an unchanging network of connections. Now, this is fine if what we are interested in is the spread of a computer virus over the internet, since the timescale for the spread of the virus is small compared with the rate at which computers are added to or removed from the network. Thus it is reasonable to treat the network as static. This is a long way from the (much more interesting) situation of a fission–fusion system, where the passing of any information (disease or otherwise) often happens on exactly the same timescale as the making and breaking of social contacts, giving rise to a strong interplay between pattern and process. Our spot-sampled networks are very likely to reveal some of the overall social structure of a population, which will in turn inform the spread of information, and this is a useful thing to know. However, to analyze the fine structure of the spread of information through such a system requires more than infrequent censuses of group co-membership.

8.3 CLOSING REMARKS

In his groundbreaking book *Sociobiology: The New Synthesis*, Wilson (1975) recognized the conceptual importance of social network structure when he highlighted it as one of the ten "qualities of sociality." The intervening thirty years or so have seen a dramatic improvement in computing power, which has made it possible to use the network approach, with its strong dependence on computer simulations (in particular randomizations), to its full potential.

Parallel developments in the social sciences and statistical physics have brought us a number of network descriptors and tests that in conjunction with the simulations make for a very powerful tool that can integrate attribute data with relational data, thus combining conventional statistics with new developments. These developments create the potential for us to learn far more about a social network than Wilson (1975) could have envisaged. Currently the application of social network theory to animal societies is in its infancy, with the majority of investigations focusing on simple pattern description, and few investigations have begun to unlock the full potential of applying social network analysis to animals.

In this book we have suggested various ways in which to explore networks. The emphasis has been on the word "explore" because so far there is no clear recipe for a definitive approach to network analysis and no consensus in the literature on which network measures are the most biologically meaningful. We are certainly not brave enough to predict which measures and methods will stand the test of time. It seems likely that some consensus will emerge through more frequent use of network theory by a larger community of researchers over the coming years. Once we have decided how best to characterize the structure of our networks, we can start to explore ways to make the networks approach a more predictive tool, how to use networks to analyze processes on a sound footing, and maybe how to use them to look at more functional issues in behavioral biology.

At the end of this book we would like to express our hope that we have interested a wider audience of behavioral biologists in the subject of social networks, and that they will join us in pushing the boundaries of this subject!

Glossary of Frequently Used Terms

association—A form of relation deduced by spatial or temporal proximity of two or more animals. Most frequently, association is assumed through co-membership of a group.

association index—Any measure of the strength of association between two animals. The simplest measure is the "association strength," which is the number of times a pair of animals associated during the data-collection period.

association matrix—A convenient mathematical representation of a network. Each row and column represents an individual node, and the entries represent relations between two nodes.

attribute data—Data that describe properties of individual nodes.

binary network—A network in which the edges carry no weight; they either exist (represented by a 1 in the association matrix) or they don't (represented by a 0).

centrality—The extent to which a given node occupies a position that is important to the structure of the network. Node betweenness and degree are measures of centrality.

clustering coefficient—A measure of the cliquishness of a network, calculated as the fraction of a node's immediate neighbors that are themselves neighbors.

community—A set of nodes that are more densely connected among themselves than they are to the rest of the network.

component—A set of nodes interconnected directly or indirectly with each other by edges.

degree—The number of edges joined to a node.

degree correlation—The tendency for nodes with high degree to be directly connected to others with high degree.

degree distribution—The fraction of the nodes of a network that have a given degree.

density—The density of a network is the proportion of actual edges out of the total number of possible edges.

directed network—A network in which the edges have a direction associated with them; unreciprocated associations or interactions lead to directed edges.

edge—A line between two nodes, representing a social interaction or association.

edge betweenness—The number of shortest paths between pairs of nodes that pass along a particular edge.

filtering—A means of including a reduced number of nodes and edges in a network. Various methods of filtering are used throughout the book.

gambit of the group—A method of constructing a network from association-based data. Two animals are assumed to be associating if they are found in the same group.

interaction—A form of relation that occurs explicitly between two animals.

leading diagonal—The terms in a square matrix that occupy the line from top left to bottom right. Elements in the leading diagonal of an association matrix represent relation of a node with itself, which is generally not considered in this book.

Monte Carlo test—A form of randomization test in which some of the structure of the original data is kept constant, while other features are scrambled.

matrix—A table representing pair-wise relations between a number of objects.

motif—A small structural feature in a network, such as a triangle of edges.

network—A collection of nodes connected by edges. A network is a graphical representation of an association matrix.

node—Each object in a network; an individual animal in this book.

node betweenness—The number of shortest paths between pairs of nodes that pass through a particular node.

path length—The number of edges on the shortest path between two nodes.

randomization test—A method of testing network structure statistically by randomizing node labels, edge labels, or some other feature of the data.

random network—Any network in which there is a random component in deciding which two nodes are connected by each edge.

regular network—A network in which every node has the same degree.

relation—A pair-wise connection between animals; edges represent relations. In this book, a relation may arise through either association or interaction.

relational data—Data that represent interactions or associations between pairs of nodes. Networks are graphical or mathematical representations of relational data sets.

shortest path—The route along edges between two nodes that contains the fewest edges.

spring embedding—A visualization algorithm that helps with the layout of a network.

trace—The sum of the terms along the leading diagonal of a matrix.

undirected network—A network in which the relations are all, or are assumed to be, un-reciprocated. The association matrix for such a network is symmetrical about the leading diagonal.

weighted network—A network in which edges are associated with a strength or frequency of association or interaction between nodes.

References

Alba, R. D. (1982). Taking stock of network analysis: A decade's results. *Research in the Sociology of Organizations* 1: 39–74.

Albert, R., and A. L. Barabási (2002). Statistical mechanics of complex networks. *Reviews of Modern Physics* 74(1): 47–97.

Arnold, S. J., and M. J. Wade (1984). On the measurement of natural and sexual selection theory. *Evolution* 38(4): 709–719.

Barabási, A., and E. Bonabeau (2003). Scale-free networks. *Scientific American* 288: 60–69.

Barabási, A. L., H. Jeong, Z. Neda, E. Ravasz, A. Schubert, and T. Vicsek (2002). Evolution of the social network of scientific collaborations. *Physica A* 311(3–4): 590–614.

Barrat, A., M. Barthélemy, R. Pastor-Satorras, and A. Vespignani (2004). The architecture of complex weighted networks. *Proceedings of the National Academy of Sciences* 101: 3747–52.

Battiston, S., G. Weisbuch, and E. Bonabeau (2003). Decision spread in the corporate board network. *Adavances in Complex Systems* 6(4): 631–44.

Battiston, S., and M. Catanzaro (2004). Statistical properties of corporate board and director networks. *European Physical Journal B* 38(2): 345–52.

Beekmans, B.W.P., H. Whitehead, R. Huele, L. Steiner, and A. G. Steenbeek (2005). Comparison of two computer-assisted photo-identification methods applied to sperm whales (*Physeter macrocephalus*). *Aquatic Mammals* 31: 243–47.

Bejder, L., D. Fletcher, and S. Bräger (1998). A method for testing association patterns of social animals. *Animal Behaviour* 56: 719–25.

Berman, C. M., K.L.R. Rasmussen, and S. J. Suomi (1997). Group size, infant development and social networks in free-ranging rhesus monkeys. *Animal Behaviour* 53: 405–21.

Bernard, H. R., P. Killworth, D. Kronenfeld, and L. Sailer (1984). The problem of informant accuracy: The validity of retrospective data. *Annual Review of Anthropology* 13: 495–517.

Bezanson, M., P. A. Garber, J. Rutherford, and A. Cleveland (2002). Patterns of subgrouping, social affiliation and social networks in Nicaraguan mantled howler monkeys (*Alouatta palliata*). *American Journal of Physical Anthropology*: Supplement 34, 44.

Boccaletti, S., V. Latora, Y. Moreno, M. Chavez, and D.-U. Hwang (2006). Complex networks: Structure and dynamics. *Physics Reports* 424: 175–308.

Bollobás, B. (1985). *Random graphs*. London: Academic Press.

Borgatti, S. P. (2002). Netdraw: Graph visualization software. Harvard: Analytic Technologies.

Borgatti, S. P., M. G. Everett, and L. C. Freeman (2002). UCINET for windows: Software for social network analysis. Harvard: Analytic Technologies.

Borgatti, S. P., K. M. Carley, and D. Krackhardt (2006). On the robustness of centrality measures under conditions of imperfect data. *Social Networks* 28: 124–36.

Bradbury, J. W., and S. L. Vehrencamp (1998). *Principles of animal communication.* Sunderland, MA: Sinauer Associates.

Bradley, B. J., D. M. Doran-Sheehy, D. Lukas, C. Boesch, and L. Vigilant (2004). Dispersed male networks in western gorillas. *Current Biology* 14(6): 510–13.

Burley, N. (1988). Wild zebra finches have band-color preferences. *Animal Behaviour* 36: 1235–37.

Burt, R. S. (1983). Studying status/role-sets using mass surveys. In *Applied network analysis: A methodological introduction,* ed. R. Burt and M. Minor. Beverly Hills, CA: Sage.

Cairns, S. J., and S. J. Schwager (1987). A comparison of association indexes. *Animal Behaviour* 35: 1454–69.

Camazine, S., J. L. Deneubourg, N. R. Franks, J. Sneyd, G. Theraulaz, and E. Bonabeau (2001). *Self-organization in biological systems.* Princeton: Princeton University Press.

Carrington, P. J., J. Scott, and S. Wasserman, Eds. (2005). *Models and methods in social network analysis.* New York: Cambridge University Press.

Chepko-Sade, B. D., K. P. Reitz, and D. S. Sade (1989). Sociometrics of *Macaca mulatta* iv: Network analysis of social structure of pre-fission group. *Social Networks* 11: 293–314.

Chilvers, B. L., and P. J. Corkeron (2002). Association patterns of bottlenose dolphins (*Tursiops aduncus*) off Point Lookout, Queensland, Australia. *Canadian Journal of Zoology-Revue Canadienne De Zoologie* 80(6): 973–79.

Clauset, A., M.E.J. Newman, and C. Moore (2004). Finding community structure in very large networks. *Physical Review E* 70(6): 066111.

Clutton-Brock, T. H., F. E. Guinness, and S. D. Albon (1982). *Red deer: Behaviour and ecology of two sexes.* Chicago: University of Chicago Press.

Connor, R. C., M. R. Heithaus, and L. M. Barre (1999). Superalliance of bottlenose dolphins. *Nature* 397: 571–72.

Corner, L.A.L., D. U. Pfeiffer, and R. S. Morris (2003). Social-network analysis of mycobacterium bovis transmission among captive brushtail possums (*Trichosurus vulpecula*). *Preventive Veterinary Medicine* 59(3): 147–67.

Corr, J. (2001). Changes in social networks over the lifespan in male and female rhesus macaques. *American Journal of Physical Anthropology*: supplement 32, 54–55.

Costenbader, E., and T. W. Valente (2003). The stability of centrality measures when networks are sampled. *Social Networks* 25: 283–307.

Croft, D. P., B. J. Arrowsmith, J. Bielby, K. Skinner, E. White, I. D. Couzin, A. E. Magurran, I. Ramnarine, and J. Krause (2003). Mechanisms underlying shoal composition in the Trinidadian guppy (*Poecilia reticulata*). *Oikos* 100: 429–38.

Croft, D. P., J. Krause, and R. James (2004a). Social networks in the guppy (*Poecilia reticulata*). *Proceedings of the Royal Society of London Biology Letters* 271: 516–19.

Croft, D. P., M. S. Botham, and J. Krause (2004b). Is sexual segregation in the guppy, *Poecilia reticulata*, consistent with the predation risk hypothesis? *Environmental Biology of Fishes* 71: 127–33.

Croft, D. P., R. James, A.J.W. Ward, M. S. Botham, D. Mawdsley, and J. Krause (2005). Assortative interactions and social networks in fish. *Oecologia* 143(2): 211–19.

Croft, D. P., R. James, P.O.R. Thomas, C. Hathaway, D. Mawdsley, K. N. Laland, and J. Krause (2006). Social structure and co-operative interactions in a wild population of guppies (*poecilia reticulata*). *Behavioral Ecology and Sociobiology* 59(5): 644–50.

Cross, P. C., J. O. Lloyd-Smith, J. A. Bowers, C. T. Hay, M. Hofmeyr, and W. M. Getz (2004). Integrating association data and disease dynamics in a social ungulate: Bovine tuberculosis in african buffalo in the Kruger National Park. *Annales Zoologici Fennici* 41(6): 879–92.

Cross, P. C., J. O. Lloyd-Smith, and W. M. Getz (2005). Disentangling association patterns in fission–fusion societies using african buffalo as an example. *Animal Behaviour* 69: 499–506.

de Nooy, W., A. Mrvar, and V. Batagelj (2005). *Exploratory social network analysis with pajek*. New York: Cambridge University Press.

Deputte, B. L., and R. Quris (1997). Socialization processes in primates: Use of multivariate analyses. 2. Influence of sex on social development of captive rhesus monkeys. *Behavioural Processes* 40(1): 85–96.

de Waal, F.B.M. (1996). Macaque social culture: Development and perpetuation of affiliative networks. *Journal of Comparative Psychology* 110(2): 147–154.

Dobson, A. (2001). *Introduction to generalized linear models*, 2nd ed. London and Boca Raton, FL: Chapman and Hall/CRC.

Dorogovtsev, S. N., and J.F.F. Mendes (2003). *Evolution of networks: From biological nets to the Internet and www*. Oxford: Oxford University Press.

Douglas, M. E., and J. A. Endler (1982). Quantitative matrix comparisons in ecological and evolutionary investigations. *Journal of Theoretical Biology* 99(4): 777–95.

Dugatkin, L. A. (1988). Do guppies play tit for tat during predator inspection visits? *Behavioural Ecology and Sociobiology* 23(6): 395–99.

Dugatkin, L. A., and D. S. Wilson (2000). Assortative interactions and the evolution of cooperation during predator inspection in guppies (*Poecilia reticulata*). *Evolutionary Ecology Research* 2(6): 761–67.

Dunne, J. A., R. J. Williams, and N. D. Martinez (2002). Network structure and biodiversity loss in food webs: Robustness increases with connectance. *Ecology Letters* 5(4): 558–67.

Durrell, J. L., I. A. Sneddon, N. E. O'Connell, and H. Whitehead (2004). Do pigs form preferential associations? *Applied Animal Behaviour Science* 89(1–2): 41–52.

Ebel, H., L. I. Mielsch, and S. Bornholdt (2002). Scale-free topology of e-mail networks. *Physical Review E* 66: 035103.

Efron, B. (1982). The jackknife, the bootstrap, and other resampling plans. *Society of Industrial and Applied Mathematics CBMS-NSF Monographs* 38.

Erdös, P., and A. Rényi (1959). On random graphs. *Publ. Math. Debrecen* 6: 290–97.

Fager, E. W. (1957). Determination and analysis of recurrent groups. *Ecology* 38: 586–95.

Faloutsos, M., P. Faloutsos, and C. Faloutsos (1999). On power-law relationships of the Internet topology. *ACM SIGCOMM Computer Communication Review* 29: 251–62.

Faust, K. (2006). Comparing social networks: Size, density, and local structure. *Metodološki zvezki* 3(2): 185–216.

Faust, K., and J. Skvoretz (2002). Comparing networks across space and time, size and species. *Sociological Methodology* 32: 267–99.

Fewell, J. H. (2003). Social insect networks. *Science* 301(5641): 1867–70.

Flack, J. C., M. Girvan, F.B.M. de Waal, and D. C. Krakauer (2006). Policing stabilizes construction of social niches in primates. *Nature* 439(7075): 426–29.

Flack J. C., and D. C. Krakauer (2006). Encoding power in communication networks. *American Naturalist* 168: 87–102.

Flack J. C., and F. de Waal (2007). Context modulates signal meaning in primate communication. *Proceedings of the National Academy of Sciences* 104: 1581–86.

Fortunato, S., and M. Barthélemy (2007). Resolution limit in community detection. *Proceedings of the National Academy of Sciences of the United States of America* 104: 36–41.

Fowler, J., L. Cohen, and P. Jarvis (1998). *Practical statistics for field biology*. Chichester, UK: John Wiley & Sons.

Frank, O. (1978). Sampling and estimation in large social networks. *Social Networks* 1(1): 91–101.

Frank, O. (1979). Estimation of population totals by use of snowball samples. In *Perspectives on social network research,* ed. P. W. Holland and S. Leinhardt. New York: Academic Press.

Freeman, L. C. (1992). The sociological concept of "group"—an empirical test of two models. *American Journal of Sociology* 98(1): 152–66.

Girvan, M., and M.E.J. Newman (2002). Community structure in social and biological networks. *Proceedings of the National Academy of Sciences of the United States of America* 99(12): 7821–26.

Good, P. (2000). *Permutation tests: A practical guide to resampling methods for testing hypotheses*. New York: Springer.

Goodman, L. A. (1961). Snowballing sampling. *Annals of Mathematical Statistics* 32: 148–70.

Granovetter, M. (1974). *Getting a job*. Cambridge, MA: Harvard University Press.

Greenacre, M. J. (1984). *Theory and applications of correspondence analysis*. New York: Academic Press.

Griffiths, S. W., and A. E. Magurran (1998). Sex and schooling behaviour in the Trinidadian guppy. *Animal Behaviour* 56: 689–93.

Guimerà, R., M. Sales-Pardo, and L.A.N. Amaral (2004). Modularity from fluctuations in random graphs and complex networks. *Physical Review E* 70: 025101.

Guimerà, R., and L.A.N. Amaral (2005). Functional cartography of complex metabolic networks. *Nature* 433(7028): 895–900.

Gupta, S., R. M. Anderson, and R. M. May (1989). Networks of sexual contacts—implications for the pattern of spread of HIV. *Aids* 3(12): 807–17.

Haccou, P., and E. Meelis (1992). *Statistical analysis of behavioral data: An approach based on time-structured models*. New York: Oxford University Press.

Hart, B. L., and L. A. Hart (1992). Reciprocal allogrooming in impala, *Aepyceros melampus*. *Animal Behaviour* 44: 1073–83.

Harvey, P., and M. Pagel (1991). *The comparative method in evolutionary biology*. Oxford: Oxford University Press.

Helfman, G. S. (1984). School fidelity in fishes—the yellow perch pattern. *Animal Behaviour* 32: 663–72.

Hemelrijk, C. K. (1990a). Models of, and tests for, reciprocity, unidirectionality and other social-interaction patterns at a group level. *Animal Behaviour* 39: 1013–29.

———— (1990b). A matrix partial correlation test used in investigations of reciprocity and other social-interaction patterns at group level. *Journal of Theoretical Biology* 143(3): 405–20.

Hinde, R. A. (1976). Interactions, relationships and social structure. *Man* 11: 1–17.

Hoare, D. J., G. D. Ruxton, J.G.J. Godin, and J. Krause (2000). The social organisation of free-ranging fish shoals. *Oikos* 89(3): 546–54.

Holland, P. W., and S. Leinhardt (1973). Structural implications of measurement error in sociometry. *Journal of Mathematical Sociology* 3(1): 85–111.

———— (1981). An exponential family of probability distributions for directed graphs. *Journal of the American Statistical Association* 76(373): 33–50.

Huisman, M., and M.A.J. van Duijn (2005). Software for social network analysis. In *Models and methods in social network analysis,* ed. P. J. Carrington, J. Scott, and S. Wasserman. Cambridge: Cambridge University Press: 270–316.

Itani J. and A. Nishimura (1973). The study of infrahuman culture in Japan. In Precultural primate behavior, ed. E.W. Menzel. Basel: Karger: 26–50.

Jasny, B. R., and L. B. Ray (2003). Life and the art of networks. *Science* 301(5641): 1863.

Katz, L., and J. H. Powell (1953). A proposed index of conformity of one sociometric measurement to another. *Psychometrika* 18: 249–56.

Kaufman, L., and P. J. Rousseeuw (1990). *Finding groups in data: An introduction to cluster analysis.* New York: John Wiley & Sons.

Kendall, M. G. (1962). *Rank correlation methods.* London: Charles Griffin.

Kirkpatrick, S., C. D. Gelatt, and M. P. Vecchi (1983). Optimization by simulated annealing. *Science* 220(4598): 671–80.

Kleinbaum, D. (1994). *Logistic regression analysis: A self-learning text.* New York: Springer-Verlag.

Knox, K. L., and D. S. Sade (1991). Social behavior of the emperor tamarin in captivity: components of agonistic display and the agonistic network. *International Journal of Primatology* 12(5): 439–80.

Kollmann, M., L. Lovdok, K. Bartholome, J. Timmer, and V. Sourjik (2005). Design principles of a bacterial signaling network. *Nature* 438(7067): 504–7.

Kossinets, G. (2006). Effects of missing data in social networks. *Social Networks* 28: 247–68.

Krause, J., and G. D. Ruxton (2002). *Living in groups.* Oxford: Oxford University Press.

Krebs, C. J. (1998). *Ecological methodology.* Menlo Park, CA: Benjamin/Cummings.

Krebs, J. R., and N. B. Davies (1996). *An introduction to behavioural ecology.* Oxford: Blackwell Science Ltd.

Kudo, H., and R.I.M. Dunbar (2001). Neocortex size and social network size in primates. *Animal Behaviour* 62: 711–22.

Lane-Petter, W. (1978). Identification of laboratory animals. In *Animal marking: Recognition marking of animals in research*, ed. B. Stonehouse. London: Macmillan: 35–39.

Latora, V., and M. Marchiori (2001). Efficient behavior of small-world networks. *Physical Review Letters* 8719(19): 198701.

Laughlin, S. B., and T. J. Sejnowski (2003). Communication in neuronal networks. *Science* 301(5641): 1870–74.

Laumann, E., P. Marsden, and D. Prensky (1983). The boundary specification problem in network analysis. In *Applied network analysis*, ed. R. Burt and M. Minor. London: Sage Publications: 18–34.

Lee, S. H., P. Kim, and H. Jeong (2006). Statistical properties of sampled networks. *Physical Review E* 73: 016102.

Lewin, K. (1951). *Field theory in the social sciences*. New York: Harper.

Lima, S. L., and P. A. Zollner (1996). Towards a behavioral ecology of ecological landscapes. *Trends in Ecology & Evolution* 11(3): 131–35.

Lusseau, D. (2003). The emergent properties of a dolphin social network. *Proceedings of the Royal Society of London Series B-Biological Science* 270(Suppl. 2): S186–S188.

Lusseau, D. (2007). Evidence for social role in a dolphin social network. *Evolutionary Ecology*: 21(3): 357–66.

Lusseau, D., K. Schneider, O. J. Boisseau, P. Haase, E. Slooten, and S. M. Dawson (2003). The bottlenose dolphin community of Doubtful Sound features a large proportion of long-lasting associations—can geographic isolation explain this unique trait? *Behavioral Ecology and Sociobiology* 54(4): 396–405.

Lusseau, D., and M.E.J. Newman (2004). Identifying the role that animals play in their social networks. *Proceedings of the Royal Society of London Series B-Biological Sciences* 271: S477–S481.

Lusseau, D., B. Wilson, P. S. Hammond, K. Grellier, J. W. Durban, K. M. Parsons, T. R. Barton, and P. M. Thompson (2006). Quantifying the influence of sociality on population structure in bottlenose dolphins. *Journal of Animal Ecology* 75: 14–24.

Lusseau, D., H. Whitehead, and S. Gero (2008). Incorporating uncertainty in the study of animal social networks. *Animal Behaviour*: in press.

MacCarthy, T., R. Seymour, and A. Pomiankowski (2003). The evolutionary potential of the *Drosophila* sex determination gene network. *Journal of Theoretical Biology* 225(4): 461–68.

Magurran, A. (2003). *Measuring biological diversity*. Oxford, UK: Blackwell Science.

Magurran, A. E. (2005). *Evolutionary ecology: The Trinidadian guppy*. Oxford, UK: Oxford University Press.

Manly, B.F.J. (1995). A note on the analysis of species co-occurrences. *Ecology* 76: 1109–15.

——— (1997). *Randomization, bootstrap and Monte Carlo methods in biology*. London: Chapman & Hall.

Manson, J. H., C. D. Navarrete, J. B. Silk, and S. Perry (2004). Time-matched grooming in female primates? New analyses from two species. *Animal Behaviour* 67: 493–500.

Mantel, N. (1967). The detection of disease clustering and a generalized regression approach. *Cancer Research* 27: 209–20.

Martin, P., and P. Bateson (2007). *Measuring behaviour: An introductory guide*. Cambridge, UK: Cambridge University Press.

Maryanski, A. R. (1987). African ape social structure: is there strength in weak ties. *Social Networks* 9(3): 191–215.

May, R. M. (2006). Network structure and the biology of populations. *Trends in Ecology and Evolution* 21(7): 394–99.

Maynard Smith, J. (1982). *Evolution and the theory of games.* Cambridge, UK: Cambridge University Press.

McDonald, D. B. (2007). Predicting fate from early connectivity in a social network. *Proceedings of the National Academy of Sciences* 104: 10910–14.

McGregor, P. K., and T. Dabelsteen (1996). Communication networks. In *Ecology and evolution of acoustic communication,* ed. D. E. Kroodsma and E. H. Miller. Ithaca, N.Y.: Cornell University Press: 409–25.

McPherson, M., L. Smith-Lovin, and J. M. Cook (2001). Birds of a feather: Homophily in social networks. *Annual Review of Sociology* 27: 415–44.

Milgram, S. (1967). The small-world problem. *Psychology Today* 2: 60–67.

Michelena, P., K. Henric, J. M. Angibault, J. Gautrais, P. Lapeyronie, R. H. Porter, J. L. Deneubourg, and R. Bon (2005). An experimental study of social attraction and spacing between the sexes in sheep. *Journal of Experimental Biology* 208: 4419–42.

Milinski, M. (1987). Tit-for-tat in sticklebacks and the evolution of cooperation. *Nature* 325(6103): 433–35.

Milo, R., S. Itzkovitz, N. Kashtan, R. Levitt, S. Shen-Orr, I. Ayzenshtat, M. Sheffer, and U. Alon (2004). Superfamilies of evolved and designed networks. *Science* 303(5663): 1538–42.

Moreno, J. (1934). *Who shall survive?* Washington DC: Nervous and Mental Diseases Publishing Company.

Moreno, Y., M. Nekovee, and A. Pacheco (2004). Dynamics of rumor spreading in complex networks. *Physical Review E* 69(6): 066130.

Nakamichi, M., and N. Koyama (2000). Intra-troop affiliative relationships of females with newborn infants in wild ring-tailed lemurs (*Lemur catta*). *American Journal of Primatology* 50: 187–203.

Newman, M.E.J. (2001a). Who is the best connected scientist? A study of scientific coauthorship networks. *Physical Review E* 64: 016131.

——— (2001b). Clustering and preferential attachment in growing networks. *Physical Review E* 64:025102.

——— (2003a). The structure and function of complex networks. *Siam Review* 45(2): 167–256.

——— (2003b). Mixing patterns in networks. *Physical Review E* 67(2): 026126.

——— (2003c). Properties of highly clustered networks. *Physical Review E* 68(2): 026121.

——— (2004). Detecting community structure in networks. *European Physical Journal B* 38(2): 321–30.

——— (2006a). Modularity and community structure in networks. *Proceedings of the National Academy of Sciences of the United States of America* 103(23): 8577–82.

——— (2006b). Finding community structure in networks using the eigenvectors of matrices. *Physical Review E* 74:036104.

Newman, M.E.J., S. H. Strogatz, and D. J. Watts (2001). Random graphs with arbitrary degree distributions and their applications. *Physical Review E* 64: 026118.

Newman, M.E.J. and M. Girvan (2004). Finding and evaluating community structure in networks. *Physical Review E* 69(2): 026113.

Noh, D. J., and H. Rieger (2002). Stability of shortest paths in complex networks with random edge weights. *Physical Review E* 66: 066127.

Nowak, M. A., and R. M. May (1992). Evolutionary games and spatial chaos. *Nature* 359(6398): 826–29.

Nowak, M. A., S. Bonhoeffer, and R. M. May (1994). Spatial games and the maintenance of cooperation. *Proceedings of the National Academy of Sciences of the United States of America* 91(11): 4877–81.

Ohtsuki, H., C. Hauert, E. Lieberman, and M. A. Nowak (2006). A simple rule for the evolution of cooperation on graphs and social networks. *Nature* 441(7092): 502–5.

Onnela, J.-K., J. Saramäki, J. Kertész, and K. Kaski (2005). Intensity and coherence of motifs in weighted complex networks. *Physical Review E* 71: 065103.

Ottensmeyer, C. A., and H. Whitehead (2003). Behavioural evidence for social units in long-finned pilot whales. *Canadian Journal of Zoology-Revue Canadienne De Zoologie* 81(8): 1327–38.

Palla, G., I. Derényi, I. Farkas, and T. Vicsek (2005). Uncovering the overlapping community structure of complex networks in nature and society. *Nature* 435: 814–18.

Pastor-Nieto, R. (2001). Grooming, kinship, and co-feeding in captive spider monkeys (*Ateles geoffroyi*). *Zoo Biology* 20(4): 293–303.

Pastor-Satorras, R., and A. Vespignani (2001). Epidemic spreading in scale-free networks. *Physical Review Letters* 86(14): 3200–3203.

Pattison, P., and S. Wasserman (1999). Logit models and logistic regressions for social networks: II. Multivariate relations. *British Journal of Mathematical & Statistical Psychology* 52: 169–93.

Pitcher, T. J. (1983). Heuristic definitions of shoaling behaviour. *Animal Behaviour* 31: 611–13.

Pitcher, T. J., D. A. Green, and A. E. Magurran (1986). Dicing with death—predator inspection behaviour in minnow shoals. *Journal of Fish Biology* 28(4): 439–48.

Potterat, J., L. Phillips-Plummer, S. Muth, R. Rothenberg, D. Woodhouse, T. Maldonaldo-Long, H. Zimmerman, and J. Muth (2002). Risk network structure in the early epidemic phase of HIV transmission in Colorado Springs. *Sexually Transmitted Infections* 78: i159–i163 Suppl. 1.

Proulx, S. R., D. E. L. Promislow, and P. C. Phillips (2005). Network thinking in ecology and evolution. *Trends in Ecology & Evolution* 20(6): 345–53.

Radicchi, F., C. Castellano, F. Cecconi, V. Loreto, and D. Parisi (2004). Defining and identifying communities in networks. *Proceedings of the National Academy of Sciences of the United States of America* 101(9): 2658–63.

Rausher, M. D., R. E. Miller, and P. Tiffin (1999). Patterns of evolutionary rate variation among genes of the anthocyanin biosynthetic pathway. *Molecular Biology and Evolution* 16(2): 266–74.

Reichardt, J., and S. Bornholdt (2004). Detecting fuzzy community structures in complex networks with a Potts model. *Physical Review Letters* 93: 218701.

Sade, D. S. (1972). Sociometrics of *Macaca mulatta*—linkages and cliques in grooming matrices. *Folia Primatologica* 18(3–4): 196–223.

Sade, D. S. (1989). Sociometrics of *Macaca mulatta*. 3. N-path centrality in grooming networks. *Social Networks* 11(3): 273–92.

Sade, D. S., M. Altmann, J. Loy, G. Hausfater, and J. A. Breuggeman (1988). Sociometrics of *Macaca mulatta*. 2. Decoupling centrality and dominance in rhesus-monkey social networks. *American Journal of Physical Anthropology* 77(4): 409–25.

Sade, D. S., and M. M. Dow (1994). Primate social networks. In *Advances in social network analysis*, ed. S. Wasserman and J. Galaskiewicz. California: Sage Publications: 152–66.

Saramäki, J., M. Kivelä, J.-P. Onnela, K. Kaski, and J. Kertész (2007). Generalizations of the clustering coefficient to weighted complex networks. *Physical Review E* 75 (2): 027105.

Schnell, G., D. Watt, and M. Douglas (1985). Statistical comparison of proximity matrices: Applications in animal behavior. *Animal Behaviour* 33: 239–53.

Scott, J. (2000). *Social network analysis: A handbook*. London: Sage Publications.

Sen, P., S. Dasgupta, A. Chatterjee, P. A. Sreeram, G. Mukherjee, and S. S. Manna (2003). Small-world properties of the Indian railway network. *Physical Review E* 67(3): 036106.

Seyfarth, R. M. (1976). Social relationships among adult female baboons. *Animal Behaviour* 24: 917–38.

——— (1980). The distribution of grooming and related behaviours among adult female vervet monkeys. *Animal Behaviour* 28: 798–813.

Shorrocks, B., and D. P. Croft (2006). Giraffe necks and networks. *Mpala News* 3: 3.

Sibbald, A. M., D. A. Elston, D.J.F. Smith, and H. W. Erhard (2005). A method for assessing the relative sociability of individuals within groups: An example with grazing sheep. *Applied Animal Behaviour Science* 91(1–2): 57–73.

Siegel, S., and N. J. Castellan (1988). *Nonparametric statistics for behavioural science*. New York: McGraw-Hill.

Sih, A., A. M. Bell, J. C. Johnson, and R. E. Ziemba (2004). Behavioral syndromes: An integrative overview. *Quarterly Review of Biology* 79(3): 241–77.

Slooten, E., S. M. Dawson, and H. Whitehead (1993). Associations among photographically identified Hectors dolphins. *Canadian Journal of Zoology-Revue Canadienne De Zoologie* 71(11): 2311–18.

Snijders, T. (1996a). Analysis of longitudinal data using the hierarchical linear model. *Quality & Quantity* 30(4): 405–26.

Snijders, T. A. B. (1996b). Stochastic actor-oriented models for network change. *Journal of Mathematical Sociology* 21(1–2): 149–172.

Snijders, T. A. B. (2005). Models for longitudinal network data. In *Models and methods in social network analysis*, ed. P. J. Carrington, J. Scott, and S. Wasserman. New York: Cambridge University Press: 215–47.

Snijders, T. A. B. and C. Baerveldt (2003). A multilevel network study of the effects of delinquent behavior on friendship evolution. *Journal of Mathematical Sociology* 27(2–3): 123–51.

Sokal, R. R., and F. J. Rohlf (1994). *Biometry: The principles and practice of statistics in biological research*. New York: W.H. Freeman and Co.

Solé, R. V., and J. M. Montoya (2001). Complexity and fragility in ecological networks. *Proceedings of the Royal Society of London Series B-Biological Sciences* 268(1480): 2039–45.

Solé, R. V., and S. Valverde (2006). Are network motifs the spandrels of cellular complexity? *Trends in Ecology & Evolution* 21(8): 419–22.

Stoinski, T. S., M. P. Hoff, and T. L. Maple (2003). Proximity patterns of female western lowland gorillas (*Gorilla gorilla gorilla*) during the six months after parturition. *American Journal of Primatology* 61(2): 61–72.

Strier, K. B., L. T. Dib, and J. E. C. Figueira (2002). Social dynamics of male muriquis (*Brachyteles arachnoides hypoxanthus*). *Behaviour* 139: 315–42.

Stumpf, M.P.H., C. Wuif, and R. M. May (2005). Subnets of scale-free networks are not scale-free: Sampling properties of networks. *Proceedings of the National Academy of Sciences* 102(12): 4221–24.

Sundaresan, S. R., I. R. Fischhoff, J. Dushoff, and D. I. Rubenstein (2007). Network metrics reveal differences in social organization between two fission-fusion species, Grevy's zebra and onager. *Oecologia* 151(1): 140–49.

Sutherland, W. (1996). *From individual behaviour to population ecology*. Oxford: Oxford University Press.

Tadic, B. (2001). Dynamics of directed graphs: The world-wide web. *Physica A* 293(1–2): 273–284.

Twigg, G. I. (1978). Marking mammals by tissue romoval. In *Animal marking: Recognition marking of animals in research*, ed. B. Stonehouse. London: Macmillan: 109–18.

von Dassow, G., E. Meir, E. M. Munro, and G. M. Odell (2000). The segment polarity network is a robust development module. *Nature* 406(6792): 188–92.

Vonhof, M. J., H. Whitehead, and M. B. Fenton (2004). Analysis of Spix's disc-winged bat association patterns and roosting home ranges reveal a novel social structure among bats. *Animal Behaviour* 68: 507–21.

Ward, A.J.W., M. S. Botham, D. J. Hoare, R. James, M. Broom, J. G. J. Godin, and J. Krause (2002). Association patterns and shoal fidelity in the three-spined stickleback. *Proceedings of the Royal Society of London Series B-Biological Sciences* 269(1508): 2451–55.

Wasserman, S., and K. Faust (1994). *Social network analysis: Methods and applications*. Cambridge, UK: Cambridge University Press.

Wasserman, S., and P. Pattison (1996). Logit models and logistic regressions for social networks. 1. An introduction to markov graphs and p. *Psychometrika* 61(3): 401–25.

Wasserman, S., and G. Robins (2005). An introduction to random graphs, dependence graphs, and p*. In *Models and methods in social network analysis*, ed. P. J. Carrington, J. Scott, and S. Wasserman. New York: Cambridge University Press.

Watts, D. J. (1999). *Small worlds: The dynamics of networks between order and randomness*. Princeton: Princeton University Press.

Watts, D. J., and S. H. Strogatz (1998). Collective dynamics of "small-world" networks. *Nature* 393(6684): 440–42.

Webb, C. R. (2005). Farm animal networks: Unraveling the contact structure of the British sheep population. *Preventive Veterinary Medicine* 68(1): 3–17.

Wey, T., D. T. Blumstein, W. Shen, and F. Jordán (2007). Social network analysis of animal behaviour: A promising tool for the study of sociality. *Animal Behaviour*: under review.

Whitehead, H. (1997). Analysing animal social structure. *Animal Behaviour* 53: 1053–67.

——— (1999). Testing association patterns of social animals. *Animal Behaviour* 57: F26–F29.

——— (2008). *Analyzing animal societies: Quantitative methods for vertebrate social analysis*. Chicago: University of Chicago Press.

Whitehead, H., L. Bejder, and A. C. Ottensmeyer (2005). Testing association patterns: Issues arising and extensions. *Animal Behaviour* 69: e1–e6.

Whitehead, H., and S. Dufault (1999). Techniques for analyzing vertebrate social structure using identified individuals: Review and recommendations. *Advances in the study of behavior, vol. 28.* San Diego: Academic Press: 33–74.

Wilkinson, G. S. (1985). The social-organization of the common vampire bat. 1. Pattern and cause of association. *Behavioral Ecology and Sociobiology* 17(2): 111–21.

Wilson, E. O. (1975). *Sociobiology: The new synthesis.* Cambridge, MA: Harvard University Press.

Wittemyer, G., I. Douglas-Hamilton, and W. M. Getz (2005). The socioecology of elephants: Analysis of the processes creating multitiered social structures. *Animal Behaviour* 69: 1357–71.

Wolf, J.B.W., D. Mawdsley, F. Trillmich, and R. James (2007). Social structure in a colonial mammal: Unravelling hidden structural layers and their foundations by network analysis. *Animal Behaviour,* 74: 1293–1302.

Xu, T., R. Chen, Y. He, and D. R. He (2004). Complex network properties of Chinese power grid. *International Journal of Modern Physics B* 18(17–19): 2599–603.

Yoon, S., S. Lee, S. Yook, and Y. Kim (2007). Statistical properties of sampled networks by random walks. *Physical Review E* 75: 046114.

Zimen, E. (1982). A wolf pack sociogram. In *Wolves of the world: Perspectives of behaviour, ecology and conservation,* ed. F. H. Harrington and P. C. Paquet. Park Ridge, N.J.: Noyes Publishers: 282–322.

Index